Farm Management Production and Resource Economics

NIPA® GENX ELECTRONIC RESOURCES & SOLUTIONS P. LTD.
New Delhi-110 034

About the Authors

Dr. Supriya is presently working as Assistant Professor in the Department of Agricultural Economics Acharya Narendra Deva University of Agriculture and Technology, Ayodhya (U.P.) She has completed her Doctorate Degree in Agricultural Economics from G.B.P.U.A&T, Pantnagar in year 2009 and qualified ARS Net in Economics and Agribusiness Management. She has also done Post Graduate Diploma in Business Management with Human Resources as specialization from SCDL, Pune and M.A. Economics from Kumaon University. She has received University Fellowship during Master's and Doctoral Program.

She has varied experience in the fields of research and teaching while working at IARI, New Delhi and Teaching at G.B.P.U.A&T, Pantnagar.

She has more than 50 Research articles published in reputed NAAS rated and other peer reviewed journals. In collaboration with Extension and Statistics Fellows she has contributed Book chapters and book in the area of Extension Education and Agricultural Statistics. She has also written many popular articles published in Hindi and English Magazines. She has also received Excellence awards by many societies in the area of Research and Teaching like Young Scientist award and Excellence in Teaching. She has taught various courses at Undergraduate and Post Graduate level. Her area of specialization is Agricultural Economics with Agricultural Statistics as minor.

Aditya Bhooshan Srivastava is presently pursuing Ph.D. in the Department of Agricultural Economics, Acharya Narendra Deva University of Agriculture & Technology, Kumarganj, Ayodhya (U.P.). He has also completed his Masters degree from Acharya Narendra Deva University of Agriculture & Technology, Kumarganj, Ayodhya in 2021. He has Young Research Award and Best thesis award by agricultural societies. He has 16 research papers in peer reviewed National and International journals. He has also written 16 popular articles and 5 book chapters. He also wrote manual cum workbook on Agricultural Finance and Cooperation.

Dr. Gaurav Sharma has more than 12 year experience in teaching and research in Agricultural Economics whereby he has worked at different positions in ICAR-Indian Agricultural Research Institute , New Delhi; G B Pant University of Agril. & Technology, Pantnager; Reliance Retail Pvt. Ltd; Rubber Research Institute of India, Kerala. Gaurav Sharma is presently Assistant Professor in the Department of Agricultural Economics at N. M. College of Agriculture, Navsari Agricultural University,

Navsari, Gujarat. Dr. Sharma received his Masters and Doctorate in Agricultural Economics from G B Pant University of Agril. & Technology, Pantnager in 2005 and 2009 respectively. Dr Sharma has published extensively in recognised national and international journal. Dr Sharma has lectured extensively on micro economics, farm resource management, Agricultural marketing and Production Economics. Dr. Sharma has also attended and received recognition at various conferences / seminars organised at national and international level.

Dr. Abdullah Mohammad Ghazi Al Khatib, Department of Banking and Insurance, Faculty of Economics, Damascus University, Syrian Arab Republic. Abdullah Mohammad Ghazi Al Khatib carries research in Financial Economics, Econometrics and Development Economics. He is Member of International Association of Scientific Research. His research area interest is: Economics, Econometrics & Finance, Mathematics, Agricultural & Biological Sciences, Environmental Sciences & Ecology, Geology, Earth & Planetary Sciences, Veterinary, and Meteorology & Atmospheric Sciences. He has worked as a reviewer for different journals. He has published many papers in peer-review journals.

Dr. Mostafa Abotaleb: His PhD was in developing artificial intelligence and machine learning models from South Ural State University, Russia. He is faculty member at South Ural State University. He then became a researcher engineer for advanced research in the department of system programming at the University of South Ural. He has many publications related to developing deep learning models and classical models to improve the forecast by minimizing the error of the forecast in Natural resource management. He has also addressed sustainability issues on global level. He has specialization in developing financial technique to minimize the risk & uncertainty in farm management. He has also expertise in developing artificial intelligence for using drone in agricultural sector.

Farm Management Production and Resource Economics

Supriya
Assistant Professor
Acharya Narendra Deva University of Agriculture & Technology
Kumarganj, Ayodhya
Uttar Pradesh

Aditya Bhooshan Srivastava
Research Scholar
Acharya Narendra Deva University of Agriculture & Technology
Kumarganj, Ayodhya
Uttar Pradesh

Gaurav Sharma
Assistant Professor
Navsari Agricultural University
Gujarat

Abdullah Mohammad Ghazi Al Khatib
Assistant Professor
Damascus University, Syrian Arab Republic

Mostafa Abotaleb
Assistant Professor
South Ural State University
Russia

NIPA® GENX ELECTRONIC RESOURCES & SOLUTIONS P. LTD.
New Delhi-110 034

NIPA® GENX ELECTRONIC RESOURCES & SOLUTIONS P. LTD.

101,103, Vikas Surya Plaza, CU Block
L.S.C.Market, Pitam Pura, New Delhi-110 034
Ph : +91 11 27341616, 27341717, 27341718
E-mail: newindiapublishingagency@gmail.com
www: www.nipabooks.com

For customer assistance, please contact
Phone: + 91-11-27 34 17 17
Fax: + 91-11-27 34 16 16
E-Mail: feedbacks@nipabooks.com

ISBN: 978-93-58875-60-7

Composed and Designed by NIPA.

Preface

World as well as Indian Agriculture is dynamic and constantly changing. Being core courses of Agricultural Economics subjects like Farm Management Production and Resource Economics assumes vital and significant importance in the agro–socioeconomic and environmental development of the country both at macro and micro level.

It is playing a catalytic role in strengthening the farm business and augmenting the productivity of scarce resources by balancing environmental condition. Farm finance can also reduce regional economic & environment imbalances, and is equally good at reducing the inter–farm asset and wealth variations.

This book **"Farm Management, Production and Resource Economics"** is about an introduction on Farm Management Production and Resource Economics concepts.

The book is having 11 chapters where concepts are nicely explained by theoretical conceptual framework along with graphs or pictures followed by objective exercises & workspace for practices.

The entire book is written in straight forward language for simple comprehension. It will suit the course for undergraduates in agricultural universities, rural universities and traditional universities. Students, professors, and researchers who work in applied economics, management of farm & natural resources etc. can use it as a reference book also.

It is passionately hoped that students will make the best use of the book to gain sufficient knowledge and practice for the recommended syllabus and for other competitive exam as well as beneficial for faculties.

Authors

Contents

Glossary

- **Advances:** It is the amount of loan advanced by an institution, during a particular period; year or season.
- **Agricultural Holding:** The total area of land owned by an individual or joint family whether cultivated by the family or rented out.
- **Agribusiness:** The sector of the economy concerned with the production, processing and distribution of agricultural products, including farm supplies (machine labour, human labour, bullock labour, fertilizers, pesticides, etc.) essential for farm production, various agricultural services (processing, storage, etc.), and also the economic agencies (financial institutions, warehouses, marketing agencies etc.) that serve agricultural producers.
- **Allocative Efficiency:** It occurs when no possible reorganization of production can make any one better off without making some one worse off.
- **Asset:** An item of value is called an asset.

or

- An asset is anything that has current or future economic value to a business.
- **Average cost:** Cost per unit of output, where the cost of all inputs are included. It is written as,

$$ATC = \frac{TC}{Output\,(X)} \text{ or } AFC + AVC$$

- **Average Fixed Cost:** It is the fixed cost per unit of output. It can also be written as,

$$AFC = \frac{TFC}{Output\,(X)}$$

- **Average Physical Product:** It is the total product divided by the total quantity of inputs being used in production process Bad debt: A debt that is difficult to collect.

- **Backward Linkage:** Refers to the relationship between an industry or firm and the suppliers of its inputs. A change in the output of the industry will get transmitted backwards to the supplier of its inputs by changing in demand for inputs.
- **Bank Credit:** Refers to the lending by the banking system by whatever means; bank advances, discounting bills or purchasing securities.
- **Bullet Loan:** A single repayment loan, having no amortization.
- **Budget constraint:** An equation that represents all combinations of goods and services that one could acquire that exactly exhaust one's budgetary resources.
- **Balance of Payments (BP):** This records the sources and uses of foreign exchange. This account records all payment transactions between the home country and the rest of the world, and is divided into two parts: the current account and the capital and financial account. Also known as: balance of payments account.
- **Balance Sheet:** A record of the assets, liabilities, and net worth of an economic actor such as a household, bank, firm, or government.
- **Capital:** It is a produced means of production. Karl Marx defined capital as a crystallised form of labour.
- **Ceteris Paribus**: Economists often simplify analysis by setting aside things that are thought to be of less importance to the question of interest. The literal meaning of the expression is 'other things equal'. In an economic model it means an analysis 'holds other things constant'.
- **Cartel:** A group of firms that collude in order to increase their joint profits.
- **Cost Minimisation:** A given level of output is produced by choosing that input combination which costs the farmer the least. Creditor: The party to whom a debt is owed. In the case of a loan, the lender is the creditor. Credit Rationing: It is one of the methods of credit control and regulation of money supply. It is a method by which the Reserve Bank of India (RBI) seeks to limit the ceiling of loans and advances and in certain cases fixes ceiling for specific categories of loans and advances. If the rationing of credit is done with reference to the total amount, it is a quantitative control, but if it is done with reference to the specific types of credit, then it is qualitative control.
- **Cost of Cultivation:** The expenditure incurred on all input and servicing in raising a crop on an unit area, is referred to as cost of cultivation. Example: Cost in Rs./ha.

- **Cost of Production:** The expenditure incurred in processing a unit quantity of all output is called cost of production. Example : Costs in Rs./ tonne or quintal.
- **Credit Squeeze:** It refers to Government restrictions on the expansion of bank loans and advances as a part of policy to check rising prices in the wake of increased aggregate demand.
- **Current Assets:** The assets that could be quickly converted into cash within a short time, usually a year e.g. agricultural produce, livestock produce, etc.
- **Current Liabilities:** The debts that must be paid in the very near future.
- **Cost Benefit Analysis:** An evaluation method that assesses the economic efficiency of policies, projects or investments by comparing their costs and benefits in present value terms. This type of analysis may include both market and non- market values and accounts for opportunity costs.
- **Cooperation:** Participating in a common project that is intended to produce mutual benefits.
- **Debenture:** It is a bond executed by a company or corporation, showing the debt owed to donor of money. This bond is not generally protected by a specific lien or mortgage on property. These bonds show a specific amount of money, i.e. principal at a specified date or over time period, during which interest is paid to the donor at a specified rate.
- **Debt Capital:** Capital obtained from borrowed funds.
- **Deed:** A deed is a document in writing or printing, which is signed, sealed and delivered by the parties thereto.
- **Discounting:** The process of calculating the present value of a stream of future values (benefits or costs). Discounting reflects individuals' or society's time preference and/or the productive use of capital. The formula for discounting or calculating present value is:
- **Dependent Variable:** A variable that is governed by another variable
- **Dividend:** The share of profits distributed to the share-holder of a company is called dividend. The dividend may be payable at a fixed rate as on preference shares, depending upon the profits that are made.
- **Discount Rate:** The rate used to determine the present value of a stream of future costs and benefits. The discount rate reflects individuals' or society's time preference and/or the productive use of capital.
- **Economies of Scale:** If the farm expands its size and streamlines its production process of crops and livestock products, its unit cost of production will be reduced considerably. Such reduction in the cost or rise

in the level of output is attributed to internal economies. The prospects of increasing returns to farm will also be possible if the farm adopts new technology, new marketing techniques or any other external factors. Such economies of scale are termed as external economies of scale. These are better explained by theory of increasing returns propounded by Adam Smith in his book The Wealth of Nations.

- **Equity Capital:** It is the capital obtained from own savings, gifts, etc.
- **Efficiency:** It means the absence of wastage or using resources as effectively as possible to satisfy the farmer's needs and goals.
- **Ecosystem:** A dynamic complex of plant, animal, and micro-organism communities and their non-living environment interacting as a functional unit.
- **Economic Activities:** All these activities that give income to the people are called Economic activities.
- **Evaluate:** To assess the overall effect of a policy or investment.
- **Farm:** It is piece of land where crop and livestock enterprises are taken up under single management and has specific boundaries.
- **Farm Management:** It means putting crop and livestock husbandry to work on the farm to obtain maximum profit.
- **Finance:** It is capital in monetary form. It is in the form of funds either borrowed or lent, normally for investment purposes through financial institutional agencies.
- **Financial Intermediaries:** All the financial institutions, viz. commercial banks, insurance companies, finance companies, investment trusts, etc., are included in the term financial intermediaries in a broad sense. But in a narrow sense, it excludes commercial banks.
- **Firm:** In economic sense, any business unit producing goods and services with ultimate objective of earning profits is termed as firm. A farm is said to be a firm, if it acquires and combines necessary resources under entrepreneurial control to produce crops and livestock products on sound business lines aiming at profit maximisation. Management of the firm may take various forms, such as proprietorship, partnership and corporation.
- **Fixed Asset:** An asset that is permanent or that will be used continuously for several years. Fixed Capital: It consists of investment in land improvements, buildings, irrigation channels, irrigation structures, machinery, livestock, etc.
- **Factors of Production**: The ingredients necessary for economic activity: land, labour, capital and entrepreneurship, which is needed to bring the other three elements together.

- **Fixed Costs:** These costs do not vary with the level of output produced by the farm. These are also known as overhead costs. Such costs remain even if no output is produced by the farm hence the name fixed costs. Rental value of land, depreciation of farm machinery, equipment, farm structures, taxes, revenue, cess, interest charges, repairs and maintenance charges, etc. constitute fixed costs. These costs are fixed in the short run, but in the long run there are, in fact, no fixed costs.
- **Fixed Inputs:** These refer to resources or factors which are used by the farm in set amount for producing crops and livestock products. Generally the size of the farm, buildings, machinery, equipment and farm structures at a given point of time whose quality is fixed, are classified as fixed inputs. In the short run these inputs are fixed, but in the long run no input is fixed. All are variable inputs, because the farm can alter its size, its operation and its output.
- **Forward Linkage:** Refers to the relationship between an industry or firm and other industries or firms, which employ its output as an input. A change in output or price will get transmitted forward to users of its product.
- **Gestation Period:** It is the period required for the investment in a project to produce a visible return.
- **Guarantor:** Though a bank may be unwilling to grant loan to one person but it may advance loan, if a second person stands as guarantor for him, i.e., he gives an undertaking, that he repays the loan in case the borrower fails to do so.
- **Hedging:** The risk of monetary loss due to price fluctuations is sometimes avoided by a marketing technique known as hedging. This technique is adopted by food grain traders, mostly wholesalers. A food grain dealer buys the product at current market prices (spot prices) and at the same time he sells an equal amount in the future market keeping some normal profit. Thus he guards himself against price decline in the future. This practice is also very common in foreign exchange market.
- **Hedonic Pricing Methods:** A method for pricing ecosystem services. Hedonic price models assume that the price of a product reflects embodied characteristics valued by some implicit or shadow price.
- **Heteroscedasticity:** An econometric problem in which the variance of the error term in a regression equation does not remain constant between observations. This problem results in the ordinary least squares procedure not producing best linear unbiased estimates.
- **Hypothecation:** In the case of hypothecation, the legal ownership of the borrowed asset, e.g., tractor, remains with lending institution, but the

physical possession is with the borrower. The loans for tangible property such as machinery and equipment are generally hypothecated loans.

- **Human Resources:** By the term human resources we mean the size of population of a country along with its efficiency. educational qualities, productivity, organizational abilities and farsightedness.
- **Income Statement:** It is a summary of income, expenses and profit or loss, from farm operations for a given period, usually a year.
- **Independent Variable:** It is variable whose value does not depend on other variable Such variable influence the dependent variable. Example Land, Labour, Liquid, Money, fertilizer, etc.
- **Insolvency:** It means the state of being unable to pay one's debts. An insolvent person or company may, after various legal processes are declared bankrupt.
- **Intangibles:** Costs or benefits which either cannot be quantified or at least cannot be priced.
- **Isocost Line:** It is a graphic line on X, Y input plane showing various possible combinations of two inputs, X and Y that can be purchased with a fixed amount of money. The slope of the isocost line represents the ratio of prices of two inputs and it is used to determine combination of two inputs to produce a given level of output with minimum cost. This means at the point of tangency of isocost line and isoquant, we get the least cost combination of two inputs to produce a given level of output.
- **Isoquant:** It is a line on the two dimensional graph showing the various possible combinations of two inputs which will yield the same level of output (unit isoquant). The slope of isoquant is negative and indicates the marginal rate of substitution of two inputs at the particular point on the curve. The convex shape of isoquant is due to diminishing marginal rate of substitution. If the two factors are perfect substitutes, the isoquant is a straight line.
- **Issue:** A risk which has already occurred is considered as issue.
- **Issued Capital:** It refers to that part of the authorised capital of a company, which has been subscribed.
- **Marginal Cost:** It is the additional cost of producing an additional unit of output

  $$MC = \frac{\Delta C}{\Delta X}$$

- **Marginal Physical Product:** It is the addition to the total output (1) as a result of using of an additional unit of input (X), keeping other factors fixed.

- **Natural Resources:** A natural resource is anything in, on, or above the earth that people use to meet their needs.
- **Production:** The process whereby some good and services called input are transferred into other goods called product or output.
- **Primary Activities:** These include those occupations which are closely related to man's natural environment. eg. Agriculture, hunting etc.
- **Revenue:** It is the change in the total revenue resulting from the disposal of an additional unit of output. In a perfectly competitive market, marginal revenue is equal to price, because of infinitely elastic demand curve.
- **Resources:** Anything that aids in production is called a resource. They physically enter the production process to transform into output. Example: Seeds, fertilizer, feeds, etc.
- **Mortgage:** It refers to a legal transfer of ownership of property from a debtor to creditor. The transfer of ownership becomes void as soon as debtor clears off the loan. Land is generally mortgaged for obtaining term loans as a security. Simple mortgage is being followed in financial institutions if the farmer-borrower has ancestral property. Equitable or registered mortgage is followed when the borrower has got self-acquired property.
- **Net Present Worth:** It results when the discounted value of the expected costs of an investment are deducted from the discounted value of the expected returns.

$$NPW = \sum_{t=0}^{T} \frac{R_t - C_t}{(1+r)^t}$$

- **Negative Externalities:** Negative externalities occur when the consumption or production of a good causes a harmful effect to a third party.
- **Nominal:** The term 'nominal' indicates that a reported value includes the effect of inflation. Prices, values, revenues etc. reported in 'nominal' terms cannot be compared directly across different time periods. See also real and constant prices.
- **Operational Holding:** Total land area held under single management for the purpose of cultivation. It excluded any land leased out to another person.
- **Optimality:** It is an ideal situation in which costs are minimum or profit is maximum.

- **Production Risk:** This is the risk to production arising from natural causes such as pest attack, extreme weather phenomenon, breakage of/technical snags in machinery at crucial time, farm fires, etc.
- **Profit:** The difference between the revenue received by a firm and the costs incurred in the production of goods and services. Value-added, profit and producer surplus are similar measures of the net benefit to producers. Although they differ slightly, the terms are used synonymously for this report to represent economic value.
- **Risk:** A situation when all possible outcomes are known for a given management decision and the probability associated with each outcome is known.
- **Revenue:** Money income that a firm receives from the sale of goods and services (often used synonymously with gross revenue).
- **Solvency:** The ability of the business farm to meet its debts or obligations to pay, is referred to as solvency. Actual solvency refers to the ability of the farm to cover all types of liabilities, but technical solvency of the farm enables it to cover its current obligations/debts.
- **Secondary Activities:** The occupations which produce finished goods by using the products of primary activities as raw material.
- **Skilled workers:** A skilled worker is one who has proper training and education to work in a particular field.
- **Tangible Assets:** The term used for physical assets such as plant and machinery, which are distinguished from intangible assets.
- **Technical Efficiency:** It is the ratio of output to input.
- **Technical Coefficient:** The amount of input per unit of output is called technical coefficient.
- **Working Capital:** It includes such items of capital as seed, feed, fertilizers, pesticides, etc. It is also called as operating or recurring capital.
- **World Bank:** An institution set up, like the International Monetary Fund, as part of the Bretton Woods agreement. The World Bank's main activity has been to provide loans and advice to developing countries, but it also conducts research into issues such as poverty and migration.
- **Welfare:** An individual's satisfaction of their wants and needs. The human satisfaction or utility generated from a good or service.

1

Farm Management An Introduction

Abstract

This chapter explores farm management, a subfield of economics that addresses decision-making on the organization and operation of a farm. The goal is to secure maximum continuous net income while ensuring the welfare of the farm family.

1.1. Meaning

- **A farm** is a piece of land, where crops and livestock enterprises are taken up **under a single management** and has specific **boundaries**.
- **Management** is the **act or art** of managing.
- Thus, **farm management** means putting the crop and livestock husbandry to work on the farm to obtain maximum profit.

1.2. Definitions

Various authors defined farm management in different ways as presented below:

1. **Farm management** is defined as the science that deals with organization and operation of the farm in the context of efficiency and continuous profits (J.N. Jefferson).
2. **Farm management** is defined as the art of managing a farm successfully as measured by the test of profitableness (Gray).
3. **Farm management** is defined as the art of applying business and scientific principles to the organization and operation of the farm (Andrew Boss).
4. **Farm management** is a branch of agricultural economics, which deals with wealth earning and wealth spending activities of farmer in relation to the organization and operation of the individual farm unit for securing the maximum possible net income (Bradford and Johnson).

5. **Farm management** is defined as the science of organization and management of the farm enterprises for the purpose of securing greatest continuous profits (G.F. Warren).
6. **Farm management** is the decision-making process whereby limited resources are allocated to a number of production alternatives to organize and operate the business in such a way to attain some objectives (Ronald D. Kay).
7. **Farm management**, as the sub-division of economics, which considers the allocation of limited resources within the individual farm, is a science of choice and decision-making and thus a field requiring studied judgment (Heady and Jensen).

Thus in simple words, farm management can be defined as **a science which deals with judicious decisions on the use of scarce farm resources, having alternative uses to obtain the maximum profit and family satisfaction on a continuous basis from the farm as a whole and under sound farming programmes**. In other words, farm management seeks to help the farmer in deciding problems like what to produce, how much to produce, how to produce and when to buy and sell and in organization and managerial problems relating to these decisions.

1.3. Importance

- Each farm unit has the capacity to produce a certain quantity of crop and livestock products. When **aggregated**, it represents the total agricultural production of a nation. Thus, the prosperity of the nation depends upon prosperity of the farmers. The prosperity of farmers depends upon managerial ability of the farmers which ultimately depends upon the knowledge of farm management.
- In the context of **increased commercialization**, there is a greater need to improve the **managerial abilities** of the farmers.
- All **farm management tools** are helpful to the farmers in solving their managerial problems for successful operation of the farm business.

1.4. Goal

The overall goal of farm management is **to manage farm businesses effectively** in order to achieve **continuous and maximum returns from the farm as a whole**.

1.5. Objectives

1. To examine production pattern and resource use on the farm.
2. To identify the factors responsible for the present production pattern.
3. To determine the conditions of optimality in the resource use and production pattern.
4. To analyze the extent of sub-optimality in the resource use.
5. To suggest ways and means in getting the present use of resources to optimality.

1.6. Farm Management Process

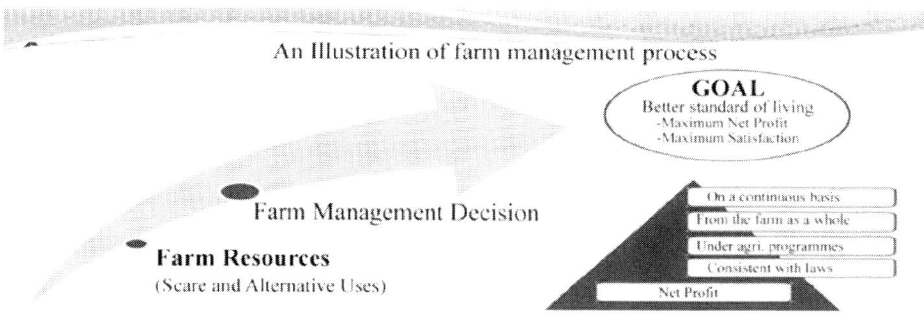

1.7. Scope (Subject Matter)

- It falls in the field of microeconomics.
- It treats every farm as a separate unit.
- It deals with the allocation of resources at the level of individual farm.In general,
- Which crops/livestock and their combination to grow?
- What amount of resources to be applied?
- How the various farm activities to be performed?

1.8. Nature and Characteristics of Farm Management

1. It is Practical Science

It aims at testing the applicability of those facts and findings and showing how to put these results to use on a given farm situation. E.g. Threshing of wheat can be done by different methods which have their own weak and strong points. **A farmer has to select a method which is more practicable and economical to his particular farm situation.**

2. It is Profitability Oriented

Agronomist and plant breeders are concerned with obtaining the **maximum yield** per acre irrespective of the profitability of inputs used but the farm management specialist always considers the cost involved in producing each unit of output and decides the **optimum level of production**.

In the decision making process, **profitability** is thus the major criterion of selection/ adoption of an enterprise or practice. In short, when other sciences deal with **physical efficiency**, the farm management concerned with **economic efficiency**.

3. It Adopts Integrated Approach

The facts and findings of other science such as weather, environment, health, etc. are coordinated for the solution of various problems of individual farmer with a view to achieving certain desired goals.

4. It is much Broader Field

It requires knowledge from many other sciences for making its own decisions. A farm management specialist has to thus a "**Jack of many trades and master of one**".

5. It has Micro Approach

Each farm unit to be studied, guided or planned **individually**.

6. Farm unit as a Whole

A farm as a **whole** is considered to be the unit for making decisions because the objective is to maximize the returns from the whole farm. Farm management considers all possible enterprises for the farm as a unit. It is much more concerned with **total crop productivity** instead of one crop productivity.

1.9. Relationship with Other Science

1. Biological Sciences

Biological sciences provide information on input-output relationship for various enterprises, which facilitate choices for the entrepreneur. Farm manager needs such information in the decisions related to production efficiency. For example,

- **Agronomy:** provides the information on physical input-output relationships in crop production.

- **Animal husbandry:** tells us the physical relationship between feed ingredients and body-weight of livestock.
- **Agricultural engineering:** explains the efficiency of various implements and machinery in performing various farm operations.

2. Economics

Farm management draws the tools and techniques from economic theory.

- The law of variable proportions
- The law of substitution
- The law of equi-marginal returns, etc.

3. Supporting Sciences

Statistics: provide **methodology** in the collection and analysis of data.

4. Social Sciences

- **Psychology:** reflects attitudes of the farmers towards decision-making on the farm reflect their.
- **Philosophy and religion:** influence the decisions of the farm managers, particularly in selection of enterprises. Religious feelings stand in the way of choosing a particular livestock enterprise like poultry, fish farming, the cultivation of narcotics etc. though it is profitable to the farmers.

5. Government Policies

Government policies also influence the production decisions.

- **A liberal export policy encourages** the farmer to go for it. On the other hand,
- **Restrictions** in the movement of food grains, supply of irrigation water, etc., **discourage** them and opt for other alternatives.

1.10. Farm Management Decisions

Farm management implies decision-making process. Several decisions need to be made by the farmer as a farm manager. They are broadly classified into three groups.

1. Organizational management decisions,
2. Administrative management decisions
3. Marketing management decisions

1. Organizational Management Decisions

They are sub-divided into operational management decisions and strategic management decisions.

i. **Operational Management Decisions:**

- These involve **less investment.**
- These are made **more frequently.**
- The effect of these decisions is **short lived.**
- Can be reversed **without incurring much cost.**
- These are also known as **tactical decisions**.

1. **What to produce?** - aims at **profit maximization**.
2. **How to produce?** - aims at **cost minimization.**
3. **How much to produce?** - aims at **resource optimization**.

ii. **Strategic Management Decisions**

- These involve **heavy investment**.
- These are made **less frequently**.
- The effect is **long lasting**.
- These decisions **cannot be altered**. Reversal **incur high cost**.
- These are also known as **basic decisions**.

1. Size of the Farm: Advantage and Disadvantage
   - Large farms enjoy low cost of production.
   - Whereas productivity is high on small farms.
2. **Use of Machinery or Labour or a combination of two.**
3. **Construction of Farm Buildings.**
   - Involves huge capital requirements.
4. **Irrigation, Conservation and Reclamation Programmes.**
   - Help in improving soil productivity.
   - Size of the farm, availability of funds, availability of ground water, etc., influence these decisions.

2. Administrative Management Decisions

Besides organizational management decisions, the farmer also makes several administrative decisions according to Government policies.

1. **Financing the Farm Business**
   - From whom to borrow
   - When to borrow and
   - How much to borrow.
2. **Supervision**
3. **Accounting**

3. Marketing Management Decisions

Marketing decisions are the most important under the changing environment of agriculture. These decisions include buying and selling.

a. **Buying:** A farmer has to decide the agency, the timing, and the quantity to be purchased.

b. **Selling**: What to sell, where to sell, whom to sell, when to sell and how to sell are the important selling decisions that are to be made by the farmer.

1.11. Agricultural Production Economics V/S Farm Management

Agricultural Production Economics	Farm Management
It is a science in which the principles of choice are applied to the use of land, capital, labour and management of resources in the farming industry.	It is a science of organization and operation of farm with a view to earn continuous and maximum profits.
It is a specialized branch of agricultural economics.	It is an integral part of agricultural production economics.
It deals with allocative efficiency of the use of resources in agriculture.	It deals with economic efficiency at the farm level.
It is an inter-farm study.	It is an intra-farm study.
It is microeconomic in its scope as it deals with the problems of farming industry.	It is microeconomic in its scope as it is concerned with the problems of individual farm.

2

Size of Farm

Abstract

This chapter examines how farm size affects farmers' welfare. Large farm owners enjoy a higher standard of living than small farm owners, which motivates their children to continue farming. The nation's scale economies also favor large farms. In India, farm size decreases with each generation due to inheritance rules. Farming does not provide enough income to ensure a decent standard of living, so the size shrinks as the number of dependents grows. These farms are not only small, but also scattered, making farming a low-paid profession.

Size of the farm influences the welfare of the farmers. Operators of large sized farms enioy better standard of living compared against their counterparts with small sized farms. Better standard of living in turn encourages their children to be in farming. This way the interest in farming makes the farmers to be committed to derive maximum production. Large farms further enjoy the scale economies as well. A country with this type of setting in agriculture regarding farm size marches ahead to produce larger quantity of farm products. In India, the law of inheritance brings down the size of farm operated by a farm family with every passing generation. Sub-division on one side reduces the size of farm but the size of holding is further reduced by fragmentation. With ever growing dependents in farming, the size of the farm had been dwindling over the years, as it is at present does not provide sufficient levels of income to ensure a reasonable standard of living. The size of farm that is operated by the farmer should doubly ensure this requirement. Apart from being small in size, these are scattered making the farming an unremunerative occupation. The various size groups are as follows:

1. Marginal — Below one hectare
2. Small — Between 1 to 2 hectares
3. Semi-medium — Between 2 to 4 hectares
4. Medium — Between 4 to 10 hectares
5. Large — 10 hectares and above

2.1. Size and Farm Returns

When the relevance of size of farm arises, we normally refer to large farms and small farms. Greater efficiencies in the use of resources are associated with the large farms than the small farms. During unfavourable years contributed by price and non-price factors both the categories of farms suffer, but the losses are clearly higher on large farms over small farms. But the large farms have the shock absorbing capacity to withstand, while small farms really suffer. However, for both the sizes of farms, favourable climatic conditions and favourable prices relative to costs bring in prosperity. This is what the best that can happen for a farming community. But this type of occurrence is not found frequently. Often it is found that given the favourable climatic conditions, favourable prices would not be in the offing. If favourable prices are found, climate may play havoc with the crop production.

2.2. Small Size Farms

Small size of holdings has come as a major hurdle in the way of introduction of new and improved technology. Smallness of holdings on one hand deters the use of mechanization and on the other, hand does not allow the use of new varieties of seeds, fertilizers and other modern inputs due to lack of purchasing power in the hands of small farmers. Those small holders of land who could not make out their subsistence from their respective farm holdings either sell or lease out their holdings and join either the army of farm laborers or migrate to urban areas in search of some work.

Advantages

1. In periods of low prices, the losses are smaller
2. The farmer can devote personal attention to every detail of farm business
3. From the viewpoint of society, the large sized farms may not be more desirable since these lead to concentration of economic power and hence anti-social. Small farms imply a wider diffusion of landed property.
4. The small farmer can withstand the period of economic depression or adverse weather conditions with relatively less loss.

Disadvantages

1. Small size of farm makes it impossible for the farmer to use the best available tools and the best methods of production.
2. Too small a farm fails to provide full time employment to the family members all the year round.

3. The marketing economies which are well reaped by large scale farmer are absent in the case of small farming. They are rarely able to offer regular supplies of their products for sale nor are they able to buy their requisites in sufficiently large quantities to the advantages. The small farmer is exposed to the exploitation of middlemen.
4. Another difficulty encountered by the small farmer is getting required credit at reasonable rate of interest.
5. Overhead charges such as depreciation, maintenance, storage etc. per unit are high.

2.3. Large Size Farms

Advantages

1. Greater output and greater net income are the features of the large farms in the light of efficient use of resources, of course not forgetting the multiplier effect during the periods of unfavourable climate and unremunerative prices.
2. Greater labour efficiency is attainted on large farms because of more productive work per man and small farms don't enjoy such a privilege, as too many work on a given size of farm.
3. Large farms put the machinery and equipment to the best advantage thereby reducing the costs per unit of output.
4. The fixed costs are lower on large sized farms.
5. In the purchase of supplies as well as disposal of products, large farms enjoy an advantage over the small farms. The bulk purchase of inputs like, seeds, fertilizers, feeds, etc. reduces the per unit costs. Similarly, sale of large volume of output brings down the marketing costs.
6. Diversification of the farming is a possibility on large farms compared to small farms, which normally specialize. The resource endowment base enables the large farms to enjoy such an advantage. Small farms specialize more in food crops as they are limited by the farm size and other resources.

Disadvantages

1. Unfavourable price environment results in greater losses on large farms over small farms due to higher investment. Similarly, they are set to loose in a bigger way in the years of poor yields.
2. When the lands owned by the large farms are relatively less fertile and the conditions of which cannot be altered, they are tended to incur greater losses, course, these are the specific cases.

3. Inefficient management of labour on large farms brings in greater disadvantage.

2.4. Factors Influencing Size of Farm

1. ***Population:*** Density of the population affects the size of the farm ie., pressure of population reduces the size of farm. In India as population is increasing the size of farm is decreasing, thereby affecting the viability of the very farm business.
2. ***Availability of Funds:*** Farmers with greater amount of funds can not only run the farms successfully, but are in a position to acquire additional land through purchase.
3. ***Climate:*** Congenial nature of climate facilitates diversification of farming which increases the size of farm business.
4. ***Managerial Abilities:*** Farmers with good management skills can operate large farms. Size of the farm is affected by the managerial skills.
5. ***Land Ceilings:*** Ceilings on land impose limits on the size of holdings as indicated by the state laws.
6. ***Availability of Irrigation Water:*** Under the irrigated conditions generally large farms can be operated. Between tank and canal irrigated farms against well irrigated farms, the former can be run economically from the point of view of pumping costs. When these two sources of irrigation are compared, relatively large areas can be put under cultivation in the areas served by canals and tanks.
7. ***Extensive and Intensive Cultivation:*** In extensive cultivation, large areas can be put under plough; while in intensive cultivation the size of farm is small %.
8. ***Technology to be Adopted:*** In land saving technology, the size of holding is small while in land augmenting technology the size of farm is larger.
9. ***Perishability and Durability of Commodities:*** Perishability of the commodities limits the area under cultivation as in the case of vegetables, while durability of the commodities encourage large scale farming.
10. ***Law of Inheritance:*** Because of the law inheritance, the size of holding declines as the given land is distributed among children.

2.5. Measures of Farm Size

The following measures are used to express the size of the farm.

1. ***Number of Acres or Hectares:*** Throughout the world, the size of farm is measured in terms of acres or hectares, which is easily understandable. But

this is not a precise measure, because it does not take into account other factors like labour and capital. This method cannot be used to compare farms situated in different agro-climatic situations. For example, a five-acre farm in Rayalaseema region of Andhra Pradesh cannot be compared against a five-acre farm in delta region in Coastal Andhra .

2. ***Labour:*** Total labour employed (man equivalents) on farm can be a measure of size of the farm. However, it is not precise as labour employment varies with the crops and technologies adopted. Besides, it ignores land and capital inputs.

3. ***Total Capital:*** Total capital invested in the form of land, farm machinery, farm buildings, livestock, etc., is also used as a measure of size of the farm. This is also not a precise measure, as it does not take into account labour and land.

4. ***Value of Production:*** This includes the gross value of all the farm products produced in a given year. This is a more accurate method of expressing the size of the farm, as it takes into account land, labour and capital resources. However, this method is also not totally free from any problem, as physical production though is same in different years, the value of production changes, because of changes in prices. Production also depends on managerial ability, which is subjective.

3

Production Function

Abstract

This chapter explores the production function, which shows how output can be maximized. Various types of production functions are used in agriculture, such as Linear, Quadratic, Cobb-Douglas, Spillman, Transcendental, etc.

3.1. Meaning

- **The functional relationship between inputs and outputs is known as production function.**
- It expresses the relationship between **quantity** of output and the quantities of various inputs used in production.
- It shows the **maximum amount of production** that can be produced from a given set of inputs in the existing state of technology or the **minimum quantities of various inputs** required to produce a given quantity.
- The **production function** can be expressed as under:

$$Y = f(X_1, X_2, X_3, \ldots, X_n).$$

Where, Y = quantity produced during a given period of time;

$X_1, X_2, X_3, \ldots, X_n$ = quantities of different factors used in production.

3.2. Definition

Production function is a technical and mathematical relationship describing the manner and extent to which a particular product depends upon the quantities of inputs or services of inputs, used at a given level of technology and in a given period of time.

3.3. Objective

- **The production function explains how the output can be optimized.**
- As we increase the level of inputs, the output also increases. So, it it is useful to know at which level of inputs, the **net returns** or the **profit is**

maximum. This level is known as **optimum level** of input-use. Production Function helps to find out the optimum level of input-use.

- Thus, main objective of production function is **resource optimization**.

3.4. Assumptions of Production Function Analysis

1. The production function is defined only for the non-negative values of inputs and outputs.
2. The production function presupposes technical efficiency. This means that every possible combination of inputs is assumed to result in maximum level of output.
3. The input- output relationship or the production function is single valued and continuous.
4. The production function is characterized by i) decreasing marginal product for all factor-product combinations; ii) decreasing rate of technical substitution between any two factors; and iii) an increasing rate of product transformation between any two products.
5. The returns to scale are assumed to be decreasing.
6. All the factors of production and products are perfectly divisible.
7. The parameters determining the firm's production function do not change over the time period considered. Also, these parameters are not allowed to be random variables.
8. The exact nature of any production function is assumed to be determined by a set of technical decisions taken by the producer.

3.5. Types of Production Function

There are two types of production function:

1. **Short-run** production function and
2. **Long-run** production function

1. Short-run production function

- It examines the relationship between **one variable factor** and output, keeping the quantities of **other factors fixed**.
- It is also called **The law of variable proportions**

  a. Single Variable Production Function

  $$Y = f(X_1 \mid X_2, X_3, \ldots, X_n).$$

Where, Y = output;

X_1 = variable input;

$X_2, X_3, \ldots, X_n$ = fixed inputs.

b. Two Variable Production Function

$Y = f(X_1, X_2 \mid X_3, X_4, \ldots, X_n)$.

Where, Y = output;

X_1 and X_2 = variable inputs;

$X_3, X_4, \ldots, X_n$ = fixed inputs.

2. Long-run production function

- In the long run, **all factors can be changed**. This production function studies the changes in output when all factors or inputs are changed.
- It is also called **Returns to scale**.
- An increase in scale means that all inputs or factors are increased in the **same proportion** i.e, double, triple, etc.

The production function can be expressed in three ways:

1. **Tabular form:** Production function can be expressed in the form of a table, where one column represents input, while another indicates the corresponding total output of the product. The two columns constitute production function.

Input (x)	Output (y)
0	2
10	5
20	11
30	18
40	25

2. **Graphical Form:** The production function can also be illustrated in the form of a graph; where horizontal axis (X axis) represents input and the vertical axis (Y axis) represents the output.

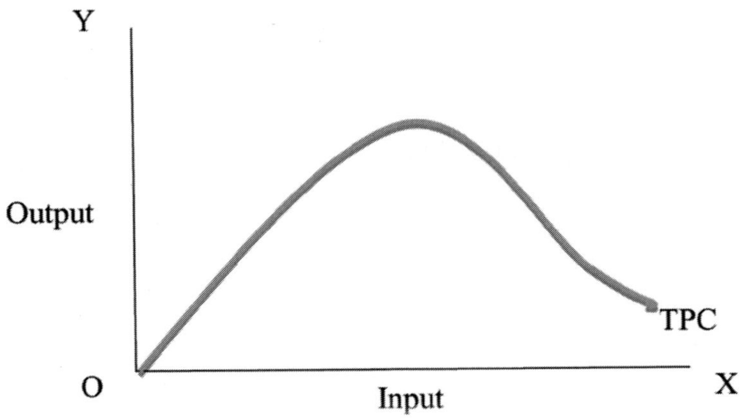

Algebraic Form : Algebraically production function can be expressed as

$Y = f(X)$

Where, Y represents dependent variable, output (yield of crop, livestock enterprise) and X represents independent variable, input (seeds, fertilizers, manure etc),

f = denotes function

When more number of inputs is involved in the production of a product, the equation is represented as

$Y=f(X_1, X_2, X_3, X_4 \ldots\ldots\ldots Xn)$

In case of single variable production function, only one variable is allowed to vary, keeping others constant, can be expressed as

$Y=f(X_1 \mid X_2, X_3 \ldots\ldots\ldots\ldots X_n)$

3.6. Several Types of Production Functions used in Agriculture are as Follows

i. **Linear Production Function:** Also known as first degree polynomial. It's algebraic form is given by

   $y = a_0 + bx$

   where,

   a_0 is the intercept and

   b is the slope of the function.

   It is not commonly used in research because it violates the basic assumptions of characteristic functional analysis.

ii. **Quadratic PF:** Also known as second degree polynomial. This type of PF allows both declining & negative marginal productivity thus embracing the second and third stage of production simultaneously.

$Y = b_0 + b_1x_1 + b_2x_2$

where b_0, b_1 & b_2, are the parameters. Such PFs are quite common in fertilizer response studies.

iii. **Cobb-Douglas PF:** The Cobb Dauglas production function is based on the empirical study of the American manufacturing industry made by Paul H. Douglas and C.W. Cobb. It is linear homogenous production function of degree on which take into account two input, labour and capital, for the entire output of the manufacturing industry. It is also known as power production function. It is most widely used in agriculture. It account for only our stage of production at at time and cannot represent constant, increasing and decreasing marginal productivity simultaneously. The Cobb Douglas Production function is expressed as:

$Q = A_0 L^{\alpha} C^{\beta}$

Where,

Q = output

A_0, α and β are positive parameter, $\alpha>0$, $\beta>0$

A_0 = total factor productivity

β = output elasticity of capital

iv. **Mitscherlich or Spill man function:** Spillman expressed the exponential relationship in the form

$Y = M\text{-}AR^{x}$

Where, Y= total output,

x = total input,

M = maximum total output which can be attained by the use of variable resources,

A = Total increase in output which can be attained by increasing x, and

R is the ratio by which successive increment are added to total production.

v. **Transcendental function:** It is also known as modify cobb Dauglas function a production function combining the characteristics of the power and exponential function is the transcendental function, The form of function is as the below:

$Y = ax_1^{b_1} x_2^{b_2} e^{b_1x_1 + b_2x_2}$

Where, Y = total output

X_1 & X_2 = Quantity of inputs

e = base of natural logarithms

$b_{1.}b_2$ = Parameter

vi. **Constant elasticity of substitution (CES) function:** An early alternative to the cobb-Douglas production function is the constant elasticity of substitution (CES) production function. While still being quite tractable, with a minimum of parameter, it is more exible than the Cobb Dauglas production Function. For the case of two input, the CES production function takes the following form.

$$Y = A\left[\delta_1 x_1^{-\sigma} + \delta_2 x_2^{-\sigma}\right]\frac{-\gamma}{\sigma}$$

Where, Y= production

X_1 and X_2 = input

A = Efficiency parameter>0

δ = Distribution parameter $0 < \delta < 1$

vii. **Square root PF:** It represents a compromise between C-D & the quadratic PF.

$$Y = a_0 + a_1\sqrt{x} + a_2 x$$

This function gets rid of the limitations of field mix of inputs for producing different levels of output inherent in the C-D production function & that of linear isoclines in quadratic function. Thus, this function allows both a diminishing TP in the same way as QF does & for declining MPs at a diminishing rate as the C-D function does.

Production function depends on the following factors

1. Quantities of inputs used
2. Technical knowledge of the producer.
3. Possible processes in production
4. Size of the firm
5. Nature of firm's organization
6. Relative prices of factors of production

4

Principles of Farm Management

Abstract

This chapter covers farm management economic principles such as factor-factor, factor-product, product-product, Equi-Marginal law, and comparative advantage law. Farm management is the application of agricultural sciences and economic principles to the organization and operation of a farm enterprise. The manager uses economic concepts to formulate goals and develop plans based on optimal resource allocation, resource substitution, and enterprise combination.

Basically, farm management is the application of agricultural sciences and economic principles to the organization and operation of a farm business. The economic principles guide the manager in setting the goals and preparing plans on the basis of optimum resource allocation, resource substitution and enterprise combination.

The various economic principles applied to farm management are discussed below:

1. Factor-Product Relationship
2. Factor- Factor Relationship
3. Product- Product Relationship
4. Law of Equi-Marginal/or Principles of Opportunity Cost
5. Law of Comparative Advantage

4.1. Factor - Product Relationship

4.1.1. Introduction

- Factor-Product Relationship is also known as **Classical Production Function**.
- It examines the relationship between **one variable factor** and output, keeping the quantities of **other factors fixed**.
- It is a **short-run production function**.
- It explains the **Law of Variable Proportion** or the **Law of Diminishing Marginal Return.**

4.1.2. Definition

The Law of Variable Proportion states **that as the proportion of one factor in a combination of factors is increased, after a point, first the marginal and then the average product of that factor will diminish**. It is, therefore, known as Law of Diminishing Marginal Return.

An increase in capital and labour applied in the cultivation of land causes in general less than proportionate increase in the amount of produce raised, unless it happens to coincide with the improvements in the arts of agriculture (*Marshall*).

If the quantity of one of productive service is increased by equal increments, with the quantity of other resource services held constant, the increments to total product may increase at first but will decrease after certain point (*Heady*).

4.1.3. Production Function Used

$Y = f(X_1 \mid X_2, X_3, \ldots, X_n)$.

Where, Y = output;

X_1 = variable input;

$X_2, X_3, \ldots, X_n$ = fixed inputs.

4.1.4. Objective

- The main objective of this relationship is to find out optimum level of resource use or **resource optimization**.
- The management problem associated with this analysis is **how much to produce**?

Limitations

The law of diminishing returns fails to operate under certain situations. They are called limitations of the law.

1. Improved methods of cultivation
2. New soils and
3. Insufficient capital.

Example

Input (X)	**Output (TP) (Y)**	**Marginal Input (ΔX)**	**Marginal Output (ΔY)**	**AP (Y/X)**	**MP (ΔY/ΔX)**	**Remarks**
0	0	-	-	-	-	Increasing at increasing rate
1	2	1	2	2.00	2	
2	5	1	3	2.50	3	
3	9	1	4	3.00	4	
4	14	1	5	3.50	5	Increasing at constant rate
5	19	1	5	3.80	5	
6	23	1	4	3.83	4	Increasing at decreasing rate
7	26	1	3	3.71	3	
8	28	1	2	3.50	2	
9	29	1	1	3.22	1	
10	29	1	0	2.90	0	
11	28	1	-1	2.54	-1	Decreasing (negative return)
12	26	1	-2	2.16	-2	

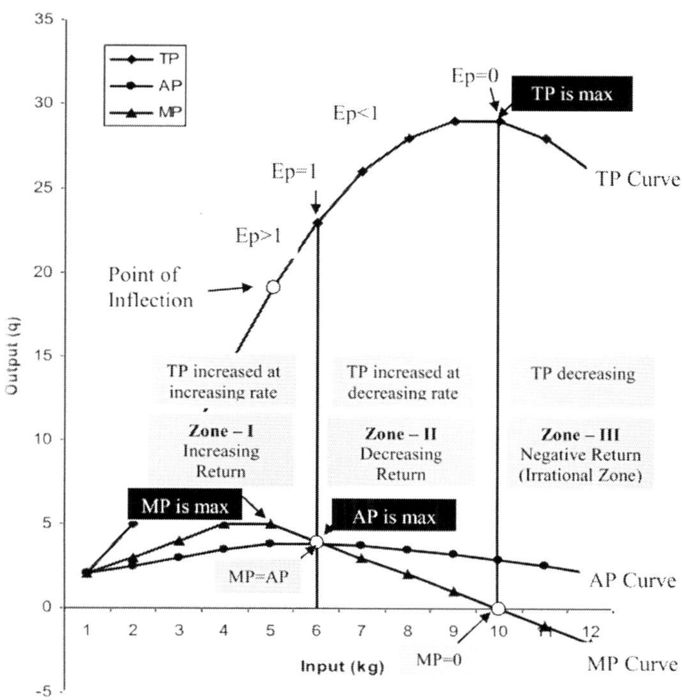

Fig. 1: Classical Production Function

Nature of TP Curve	Nature of MP Curve	Nature of AP Curve
• TP increasing at increasing rate • TP increasing at decreasing rate • TP riches at maximum and • remains constant • TP decreasing	• MP increasing • MP attains maximum • MP remains constant • MP decreasing	• AP increasing • AP attains maximum • AP is decreasing

A. Relationship between TP and MP

- When TP increasing at increasing rate, MP also increasing.
- When TP increasing at decreasing rate, MP also decreasing.
- When TP is maximum or remain constant, MP = 0
- When TP decreasing, MP becomes negative.
- When TP changes its curvature (from increasing rate to decreasing rate), MP is maximum. This point is known as point of inflection.

B. Relation between MP and AP

- When MPP is increasing, APP is also increasing. So long as MPP is above APP, the APP keeps increasing.
- When MPP curve goes below APP curve, APP starts declining, that is, when AP is decreasing the MP is always less than APP.
- When MP = AP, AP will be at maximum. Here MP curve must intersect AP curve from above at its highest point.

  So when MP > AP AP↑

  MP < AP AP↓

  MP = AP AP is at maximum.

Relationship between MP and TP	Relationship between MP and AP
• MP > 0, TP increasing • MP = 0, TP maximum • MP < 0, TP decreasing	• When MP > AP, AP increasing; Ep >1 • When MP = AP, AP is maximum; Ep = 1 • When MP < AP, AP decreasing; Ep < 1

4.1.5. Three Zones of Production Function

The classical production function can be divided into three zones.

Zone-I: Stage of increasing returns

- This region starts from the point of origin and ends where the average product reaches its highest (maximum) or where MP=AP.

- During this stage, the TP, AP and MP are increasing.
- It is notable that the MP in this stage increases but in a later part it starts declining. Though MP declining, it is greater than the AP.
- AP increases throughout this region indicating that the efficiency of all the variable inputs keeps on increasing.

Zone-II: Stage of decreasing returns

- This region starts where 1st zone ends. i.e. MP=AP and extends to the point where MP=0 or TP is maximum.
- In this zone, the TP continues to increase but at a diminishing rate.
- The MP and the AP are declining but are positive.

Zone-III: Stage of negative returns

- This region starts when MP crosses zero and becomes negative.
- In this zone the MP becomes negative.
- The TP and the AP are declining.

Rational / Irrational Zones

Zone-III

- This is a **irrational zone** because in this zone the TP is decreasing and MP becomes negative which indicates that additional quantities of inputs reduces the total output.
- So, it is not advisable to operate in this region even if the additional quantities of inputs are available free of cost.
- If a farmer operates in this zone, he will have **double loss**: i.) reduced production and ii) unnecessarily additional cost of inputs.
- For example, many farmers in **canal area** operate in this zone. They use excess water which ultimately reduces the profit.

Zone-I

- This is also **irrational zone** because, the AP increases throughout this zone indicating that the efficiency of all the variable inputs keeps on increasing. So, farmer should not stop in this zone and he must produce up to the level where AP is maximum.
- Input-use should be continued until zone-II. Hence, **it is not reasonable to stop using an input** when its efficiency is increasing. If he stops in this region, some of his resource will remain **unused or underutilized**.

- For example, most of the tribal farmers operate in this region because they don't have sufficient inputs.

Zone-II

- This is a **rational zone.**
- In this zone, TP is increasing; MP is decreasing but remains positive.
- Within the boundaries of this region is the area of economic relevance.
- Optimum use of input is somewhere in this zone which, however, can be located only when input and output prices are known.
- This zone represents the range of rational production decisions.

Summary

Zone - I	Zone - II	Zone – III
From origin to AP=MP	From AP=MP to MP=0	From MP=0 to onwards
Ep > 1	Ep < 1	Ep < 0
Irrational zone	**Rational zone**	**Irrational zone**
Stage of **increasing return**	Stage of **decreasing return**	Stage of **negative return**
TP, AP & MP all are increasing	TP is increasing but at a decreasing rate. MP & AP decreasing but are positive.	TP, AP & MP all are decreasing. MP becomes negative.
Zone of **underutilization** of resources	Zone of **optimum** level of resources	Zone of **overutilization** of resources

4.1.6. Elasticity of Production

The elasticity of production is a concept that measures the degree of responsiveness between output and input. It is independent of the units of measurement.

$$Ep = \frac{\%\ \text{change in output}}{\%\ \text{change in input}}$$

$$Ep = \frac{\Delta Y / Y}{\Delta X / X} = \frac{\Delta Y / \Delta X}{Y / X} = \frac{MPP}{APP}$$

Ep >1 in Stage I

- ≤ Ep ≤ 1 stage II

Ep is negative stage III

The point of diminishing returns can be defined to occur when MPP =APP that is Ep = 1 (lower boundary of stage II) & this is the minimum amount of variable input that will be used & it occurs when the efficiency of variable input is at its maximum. At the other end, MPP is zero, therefore Ep = 0. Thus the relevant production zone is when $0 \leq Ep \leq 1$.

The optimum level of variable input is given by:

$\Delta X_1.PX_1 = \Delta Y.\ Py$ or $\Delta Y/\ \Delta X_1. = PX_1/\ Py$
Marginal cost = Marginal revenue.

4.1.7. Optimum Level of Input Use

Optimum level of input use will be at the stage where the cost of additional input (MC) and the value of additional product (MR) are equal i.e., MC = MR. To calculate optimum level, data regarding price of input as well as output will be required.

Suppose, price of input is **Rs. 5/unit** and price of output is **Rs. 4/unit.**

Input (X)	Output (Y)	Total Cost	Total Return	Marginal Cost	Marginal Return	Net Return
0	0	0	0	-	-	0
1	2	5	8	5	8	3
2	5	10	20	5	12	10
3	9	15	36	5	16	21
4	14	20	56	5	20	36
5	19	25	76	5	20	51
6	23	30	92	5	16	62
7	26	35	104	5	12	69
8	**28**	**40**	**112**	**5**	**8**	**72**
9	**29**	**45**	**116**	**5**	**4**	**71**
10	29	50	116	5	0	66
11	28	55	112	5	-4	57
12	26	60	104	5	-8	44

In this example, the optimum use of resource is somewhere between 8 and 9 unit.

4.1.8. Application of the Law

- The law is widely applicable to **agriculture**.
- Besides, agriculture, it also applies to **mining, fisheries** and also to **building industries** where **the nature has supremacy**.
- It is therefore, often remarked that the part that the **nature** plays in

production corresponds to **diminishing returns** and the part which **man** plays conforms to the law of **increasing returns**.

- Thus, we can say that **agriculture,** where nature is supreme, is subject to **diminishing returns**, while **industry**, where man is supreme, is subject to **increasing returns**.

Reasons why agriculture is subject to the law of diminishing returns

1. Agricultural is related to nature. Man is not a complete master of nature. The operations are likely to be interrupted frequently by rain and other climatic changes.
2. The agricultural operations are spread out over a wide area, and consequently supervision cannot be very effective.
3. Scope for the use of specialized machinery is also very limited.
4. No scope for division of labour.

Reasons why industry is subject to the law of increasing returns

1. In case of manufacturing industries, **man** has the fullest scope to show his ability.
2. By the introduction of **division of labour** and the use of the most **modern technology**, production can be greatly increased.
3. Concentration of workers under one roof renders **supervision easy and effective**.

But, if the industry is **expanded too much**, supervision will become difficult and the costs will go up. In such situation, **law of diminishing returns will set in**. A prudent industrialist may not allow that stage to come at all.

Hence, we can say that the law of diminishing returns has wide applications. In agriculture it sets in earlier, while in industry, it sets in a later stage.

4.2. Factor - Factor Relationship (Law of Factor Substitution)

4.2.1. Introduction

- Any production activity requires **different inputs** to produce a given quantity of output. There are many ways of **combining these resources** in production process.
- Farm production facilitates the **substitution of resources**. For example, a farmer producing crop does make a choice between **manures and chemical fertilizers, human labour and machinery**, etc. Similarly, a dairy farmer has to decide upon the quantity of **fodder and concentrate** to arrive at feed ration.

- The factor-factor relationship is explained by the **law of factor substitution**.
- In this relationship, two factors or inputs are varied and the **output** is kept **constant**.

4.2.2. Production Function Used

$Y = f(X_1, X_2 \mid X_3, X_4, \ldots, X_n)$

Where, Y = a fixed level of output;

X_1, X_2 = variable inputs.

$X_3, X_4, \ldots, X_n$ = inputs which remains constant.

4.2.3. Objective

- The **objective** here is to find out the most appropriate combination of inputs that **costs the least** amount in producing a given level of output or **cost minimization**.
- The **management decision** associated with this analysis is **how to produce**? i.e. which combination of two factors to be used?

4.2.4. ISO-Quant

- **ISO** means equal and **quant** means quantity.
- **The curve representing all combinations of X_1 and X_2 that produce the equal quantity of output is called an iso-quant**.
- It is also termed as iso-product curve or product indifference curve.

Example: The output of paddy amounting to 1000 kg/acre can be produced using the different combinations of two inputs viz., Manure (X_1) and Urea (X_2) as under.

Combination	Manure (X_1) (qty. in cart-load/acre)	Urea (X_2) (qty. in kg/acre)
A	1	200
B	2	125
C	3	80
D	4	50
E	5	30
F	6	17
G	7	10

From the above table, it is clear that all combinations with different quantities of X_1 and X_2 result in the same level of production. **Iso-quant (Fig. 2)** represents all combinations of X_1 and X_2 that produce 1000 kg of paddy per acre.

Iso-quant can be shown for any level of output. As such, several iso-quants can be shown for various levels of output with different levels of inputs. **If a number of iso-quants are drawn in one graph, it is known as iso-quant map (Fig. 3).**

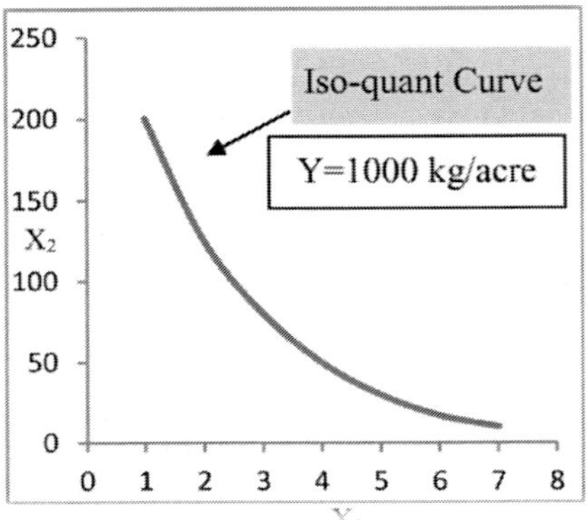

Fig. 2: Iso-quant Curve

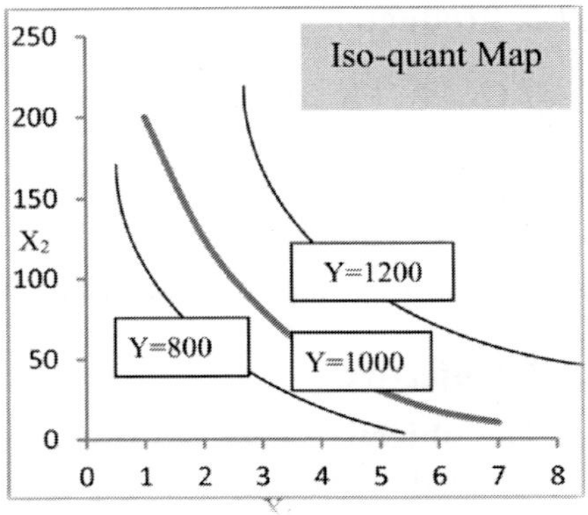

Fig. 3: Iso-quant Map

Characteristics or Properties of an iso-quant

- Iso-quants are non-intersecting.
- Iso-quant is downward sloping from left to right i.e. it is negatively sloped.
- Iso-quant placed above represent higher output and *vice-versa.*
- Iso-quant is convex to the origin.

4.2.5. Marginal Rate of Technical Substitution (MRTS)

- It is the **rate of exchange between two inputs** which are equally preferred.
- It indicates the absolute amount by which **one input is decreased or replaced to add a unit of another input.**
- Alternatively, **the quantity of one input to be sacrificed or given up in order to gain another input by one unit**, in process of substitution.

$$\text{MRTS} = \frac{\text{Quantity of replaced input}}{\text{Quantity of added input}}$$

MRTS of X_1 for X_2 is written as : $\text{MRTS}_{X1X2} = \Delta X2 \div \Delta X1$; and

MRTS of X_2 for X_1 is written as : $\text{MRTS}_{X2X1} = \Delta X1 \div \Delta X2$

The computation of MRTS is presented in following table.

Combination	**Manure (X1) (cart-load/acre)**	**Urea (X2) (kg/acre)**	**ΔX1**	**ΔX2**	**$\text{MRTS}_{X1X2} = \Delta X2 \div X1$**	**$\text{MRTS}_{X2X1} = \Delta X1 \div \Delta X2$**

A	1	200	-	-	-	-
B	2	125	1	75	75	0.013
C	3	80	1	45	45	0.022
D	4	50	1	30	30	0.033
E	5	30	1	20	20	0.050
F	6	17	1	13	13	0.077
G	7	10	1	7	7	0.143

4.2.6. Iso-Cost Line

- **Iso-cost line** represents various combinations of two inputs that can be purchased with the given amount of money.
- It is known as **price line** or **budget line**.
- It can be drawn by locating the end points of X1 and X2 that can be purchased from the given amount of money.
- It plays an important role in **determining the combination of factors** that the farmer should choose for production.

Example

Suppose, a farmer has a fund of Rs. 1000 and he has to spend on two inputs viz., X1 (Manure) and X2 (Urea). Price of X1 is Rs. 200/cart-load and that of X2 is Rs. 5/kg. Given the prices of X1 and X2, he can purchase a maximum 5 cart-load of Manure or 200 kg of Urea.

To draw **Iso-cost line** we need two extreme points X1 and X2 which can be obtained as under:

X_1 (cart-load)	X_2 (kg)
5	0
0	200

When these two points X_1 (5,0) and X_2 (0,200) are joined, we get **iso-cost line** (**Fig. 4**). Similarly, iso-cost line for budget of Rs. 750 and Rs. 1250 can be drawn as under. They will be parallel to each other.

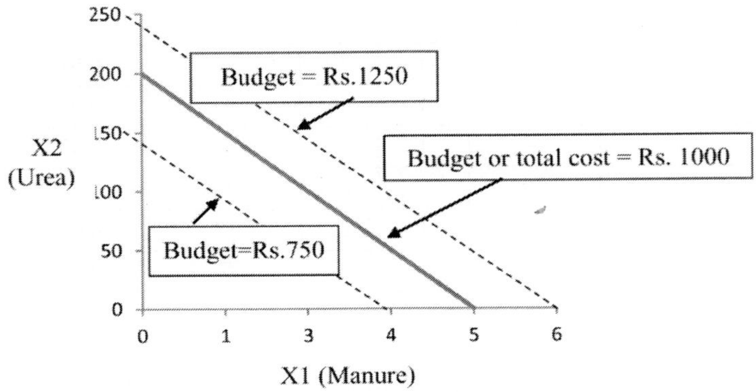

Fig. 4: Iso-Cost Lines

Characteristics of Iso-cost line

- Iso-cost line is always a **straight line.**
- Change in input prices will **change the slop** of the iso-cost line.
- When the **total outlay or budget is increased**, the iso-cost line **shifts upwards** to the right and it **moves farther** away from the origin and *vice-versa.*
- Iso-cost lines are **parallel to one other**, since relative price ratio remains constant.
- The **slop** of iso-cost line **shows the inverse price ratio** of two inputs.

4.2.7. Least Cost Combination

- Factor-factor analysis is mainly concerned with the determination of least cost combination of resources.
- There will be many combinations of two resources that produce the same level of output.
- The problem here is to find out particular combination of inputs, which produce a given quantity of output with **minimum cost**.

There are **three methods** of finding out least cost combination viz., Tabular Method, Algebraic Method and Graphic Method.

Example

A farmer wants to produce 1000 kg of paddy per acre of land. He has two types of fertilizers viz., Manures and Urea that can be used to produce the crop. The following different combinations of Manures and Urea produce the same quantity of output i.e., 1000 kg of paddy per acre.

Combination	Manure (X_1) (qty. in cart-load/acre)	Urea (X_2) (qty. in kg/acre)
A	1	200
B	2	125
C	3	80
D	4	50
E	5	30
F	6	17
G	7	10

Price of Manure is Rs. 200/cart-load and price of Urea is Rs. 5/kg.

Find out **least cost combination** using all three methods and give your advice to the farmer about how much quantity of Manure and Urea he should use to produce the required output.

Solution

Tabular Method

Combination	Manure (X_1) (cart-load/acre)	Urea (X_2) (kg/acre)	Cost of X_1 ($X1*PX_1$)	Cost of X_2 ($X2*PX_2$)	Total Cost
A	1	200	200	1000	1200
B	2	125	400	625	1025
C	3	80	600	400	1000
D	4	50	800	250	1050
E	5	30	1000	150	1150
F	6	17	1200	85	1285
G	7	10	1400	50	1450

Conclusion: Total Cost of "**C**" combination is the minimum (least) among all. Therefore, the farmer should use **3** cart-load of Manure and **80** kg of Urea to produce the given output.

Algebraic Method

Step-I : Compute the Marginal Rate of Technical Substitution (MRTS)

$$MRTS_{X1X2} = \frac{\Delta X_2}{\Delta X_1}$$

Step-II : Compute the Price Ratio (inverse)

$$PR_{X2X1} = \frac{PX_1}{PX_2}$$

Step-III : Find out least cost combination by equating MRTS with inverse PR

Calculation of MRTS and Price Ratio

Price of X1 = Rs. 200/cart-load, Price of X2 = Rs. 5/kg

Combi-nation	**X_1 (cart-load/ acre)**	**X_2 (kg/acre)**	**ΔX_1**	**ΔX_2**	**$MRTS_{X1X2} = \frac{\Delta X_2}{\Delta X_1}$**	**$PR_{X2X1} = \frac{PX_1}{PX_2}$**
A	1	200	-	-	-	-
B	2	125	1	75	75	40
C	**3**	**80**	**1**	**45**	**45**	**40**
D	4	50	1	30	30	40
E	5	30	1	20	20	40
F	6	17	1	13	13	40
G	7	10	1	7	7	40

or

Combination	**X_1 (cart-load/ acre)**	**X_2 (kg/acre)**	**ΔX_1**	**ΔX_2**	**$MRTS_{X2X1} = \frac{\Delta X_1}{\Delta X_2}$**	**$PR_{X1X2} = \frac{PX_2}{PX_1}$**
A	1	200	-	-	-	-
B	2	125	1	75	0.013	0.025
C	3	80	1	45	0.022	0.025
D	4	50	1	30	0.033	0.025
E	5	30	1	20	0.050	0.025
F	6	17	1	13	0.077	0.025
G	7	10	1	7	0.143	0.025

Conclusion

At Combination **"C", MRTS=PR** is observed. So, this combination is the least cost combination. Therefore, the farmer should use **3** cart-load of Manure and **80** kg of Urea to produce the given output.

Graphical Method

Step-I : Draw **Iso-quant** on a graph paper.

Plot X_1 (added quantity) on X axis and X_2 (replaced quantity) on Y axis to get Iso-quant.

Step-II : Draw **Iso-cost Line** on the same graph paper.

To draw Iso-cost line we need two extreme points X_1 and X_2. Suppose, we have a budget of Rs. 1000. The Maximum quantity of X_1 and X_2 can be purchased as under:

X1 (cart-load)	**X2 (kg)**
5	0
0	200

Plot these two extreme points of X1 (5,0) and X2 (0,200) on the same graph. Join these two points X_1 and X_2. This line is called Iso-Cost Line.

Step-III : If required, move the Iso-cost line or draw the parallel line in such a way that **it intersects** the Iso-quant **at one point**. This point indicates the **least cost combination** of two inputs.

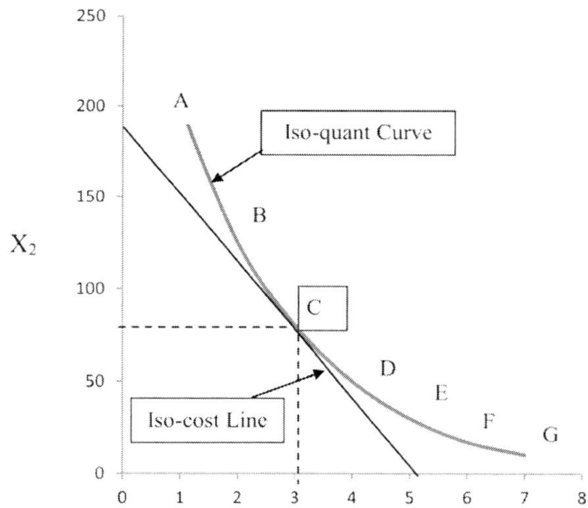

Conclusion

Combination "**C**" is found to be least cost because at this level Iso-quant curve and Iso-cost line **intersects** to each other. Therefore, the farmer should use **3** cart-load of Manure and **80** kg of Urea to produce the given output.

4.3. Product - Product Relationship (Law of Product Substitution)

4.3.1. Introduction

The basic resources of farming viz., land, labour and capital are scare. However, these scare resources have many alternative uses. These can be used in producing different crops and livestock enterprise. Farmers have to decide whether to produce crops alone or livestock alone or their combinations. The farmer should choose such a **combination** of crops and livestock enterprise that **maximizes profit**.

- Product-product relationship deals with the **allocation of resources** among different crop and livestock enterprise.
- It is explained by the **law of product substitution**.
- In this relationship, resources or **inputs are kept constant** and **products are varied.**

4.3.2. Production Function Used

$Y1 = f(Y_2 \mid X_1, X_2, \ldots, X_n)$.

Where, Y_1 and Y_2 are variable output of two products;

$X_1, X_2, \ldots, X_n$ are fixed inputs.

4.3.3. Objective

- The objective of product-product relationship is **profit maximization.**
- The management problem associated with this analysis is **what to produce**.

1. Production Possibility Curve

- It represents **all possible combinations of two products** (Y_1 and Y_2) that could be produced with **given amount of resources**.
- It is known as **opportunity curve** because it represents all the production opportunities available with a given amount of resources.
- It is also called a **Production Frontier** because the limited resources cannot help to produce anything beyond production possibility curve. It demarcates what is possible at a given quantity of inputs.

Suppose a farmer has a limited input i.e., 5 acres of land. He has two alternatives i.e., the production of Maize (Y_1) crop and/or Bajra (Y_2) crop. Problem here is how to allocate his 5 acres between Y1 and Y_2. Here, farmer can allocate entire 5 acres to Y_1 alone or Y_2 alone. In between these two extreme possibilities, he may have different options like, 1 acre to Y_1 and 4 acres to Y_2 or 2 acres to Y_1 and 3 acres to Y_2 and so on. For example, different levels of land and the corresponding levels of output of Y_1 and Y_2 are as under.

Land (acre)	Production (qtl)	
	Maize (Y_1)	Bajra (Y_2)
1	9	8
2	15	14
3	20	20
4	23	24
5	25	27

Here, possible allocation of land among two crops Y_1 and Y_2 can be done as under.

Combination	Allocation of Land in acres		Output (Production) in qtls	
	Maize (Y_1)	Bajra (Y_2)	Maize (Y_1)	Bajra (Y_2)
A	0	5	0	27
B	1	4	9	24
C	2	3	15	20
D	3	2	20	14
E	4	1	23	8
F	5	0	25	0

The **production possibility curve** for the above situation can be prepared by presenting the outputs of Y_1 and Y_2 in graphical form (**Fig. 5**).

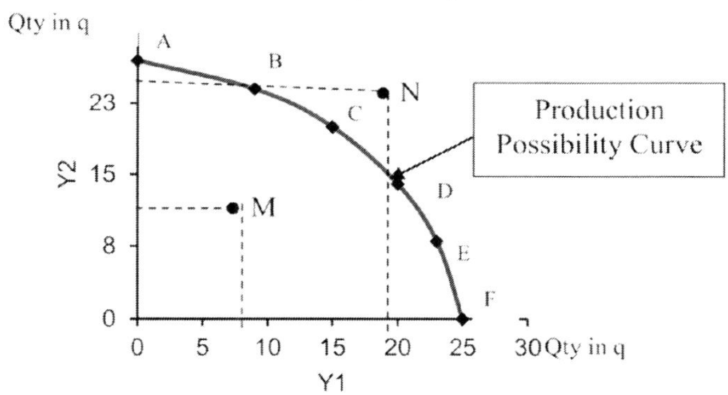

Fig. 5. Production Possibility Curve

Characteristics of Production Possibility Curve

- It is **concave** to the origin.
- The slope of the curve indicates the marginal rate of product substitution.
- Any combination of Y_1 and Y_2 inside the curve (Point M) is possible but it reflects inefficient use of resources.

- Any combination of Y_1 and Y_2 on the curve (Point A to F) is possible to produce and it reflects efficient use of resources.
- Any combination of Y_1 and Y_2 outside the curve (Point N) is not possible to produce at given amount of resources.
- Change in input levels, shifts the production possibility curve.

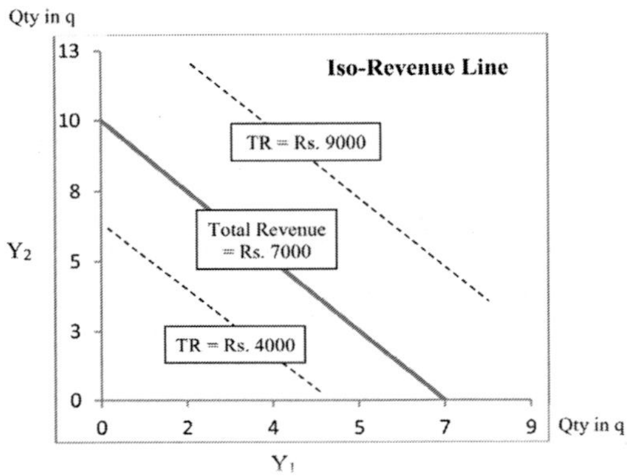

Fig. 6: Shifts in Production Possibility Curve

4.3.4. Marginal Rate of Product Substitution (MRPS)

- The absolute amount, by which **one product is decreased**, in order to **gain another** product **by a unit** is called marginal rate of product substitution.
- It is the **rate of exchange between two products** which are equally preferred.

$$\text{MRTS} = \frac{\text{Number ofunits of replaced product}}{\text{Number of units of added product}}$$

$$\text{MRPS of } Y_1 \text{for } Y_2 \left(\text{MRPS}_{Y1Y2}\right) = \frac{\Delta Y_2}{\Delta Y_1}$$

OR

$$\text{MRPS of } Y_2 \text{for } Y_1 \left(\text{MRPS}_{Y2Y1}\right) = \frac{\Delta Y_1}{\Delta Y_2}$$

4.3.5. ISO-Revenue Line

It is line, which represents all possible combinations of two products which would yield equal revenue.

Suppose we wish to obtain a total revenue of Rs. 7000. If Price of Y1 is Rs.1000/q and price of Y_2 is Rs.700/q, the expected revenue of Rs.7000 can be earned by producing 7 quintals of Y_1 or 10 quintals of Y2.

Y_1 (q)	Y_2 (q)
7	0
0	10

Plot these two extreme points Y_1 (7,0) and Y_2 (0,10). Join these two points and the line that we get is called iso-revenue line (**Fig. 7**). Similarly, the iso-revenue line for Rs. 4000 and Rs. 9000 can be drawn as under. They will be parallel to each other.

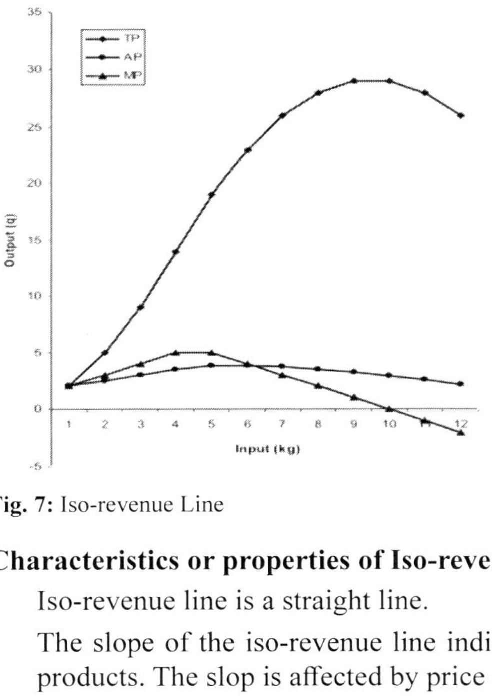

Fig. 7: Iso-revenue Line

Characteristics or properties of Iso-revenue Line

- Iso-revenue line is a straight line.
- The slope of the iso-revenue line indicates the inverse price ratio of the products. The slop is affected by price change.
- The iso-revenue lines are parallel to each other, since price ratio remains constant.
- As the total revenue increases, the iso-revenue shifts upwards and moves away from the origin and vice-versa.

4.3.6. Profit Maximization

The objective of product-product relationship is to determine which out of all possible combinations is the most profitable. For this, we require the prices of Y_1 and Y_2 products. Suppose, the price of Y_1 is Rs.1000/q and price of Y_2 is Rs.700/q, for the given example, find out the combination of two crops which gives the maximum profit.

Solution

There are three methods to find out the best combination that gives maximum profit viz., **Tabular Method, Algebraic Method and Graphical Method.**

1. Tabular Method

Combi-nation	Allocation of land (acre)		Production (qtl/acre)		Income		Total Income
	Y_1	Y_2	Y_1	Y_2	$Y_1 \times PY_1$	$Y_2 \times PY_2$	
A	0	5	0	27	0	18900	18900
B	1	4	9	24	9000	16800	25800
C	2	3	15	20	15000	14000	29000
D	**3**	**2**	**20**	**14**	**20000**	**9800**	**29800**
E	4	1	23	8	23000	5600	28600
F	5	0	25	0	25000	0	25000

Conclusion : Combination **"D"** gives highest Income. Therefore, out of 5 acres, farmershould grow **Y_1 crop in 3 acres and Y_2 crop in 2 acres of land**.

2. Algebraic Method

Step 1: Find out Marginal Rate of Product Substitution (MRPS)

Step 2: Find out inverse 'Price Ratio' for given prices of Y_1 & Y_2

Step 3: Compare both MRPS and PR ratio. MRPS=PR gives maximum returns.

Combi nation	Allocation of land (acre)		Production (qtl/acre)		ΔY_1	ΔY_2	MRPS Y_1Y_2	PR Y_2Y_1	MRPS Y_2Y_1	PR Y_1Y_2
	Y_1	Y_2	Y_1	Y_2						
A	0	5	0	27	-	-	-	-	-	-
B	1	4	9	24	9	3	0.33	1.43	3.00	0.70
C	2	3	15	20	6	4	0.66	1.43	1.50	0.70
D	**3**	**2**	**20**	**14**	**5**	**6**	**1.20**	**1.43**	**0.83**	**0.70**
E	4	1	23	8	3	6	3.00	1.43	0.50	0.70
F	5	0	25	0	2	8	4.00	1.43	0.25	0.70

Conclusion

Combination D represents **MRPS=PR**. Therefore, out of 5 acres, farmer should grow **Y_1 crop in 3 acres and Y_2 crop in 2 acres of land**.

Graphical Method

Step 1: Prepare a Production Possibility Curve

Plot Y_1 (added quantity) on X axis and Y_2 (replaced quantity) on Y axis to get production possibility curve.

Step 2 : Draw Iso-Revenue Line or Price Line using price data

To draw iso-revenue line we need two extreme points of Y1 and Y2 which can be obtained as under:

Suppose, we want to have a total revenue of Rs. 21,000. It is possible as under:

Y_1 (q)	Y_2 (q)
21	0
0	30

Plot these two extreme points of Y_1 (21,0) and Y_2 (0,30) on the same graph. Join these two points Y_1Y_2. This line is called Iso-Revenue Line.

Step 3 : Move Iso-revenue line to intersect production possibility Curve at one point

At the point where Production possibility Curve and Iso-revenue Line intersect is the point of maximum profit. At this point combination of two products Y1 & Y2 gives maximum profit.

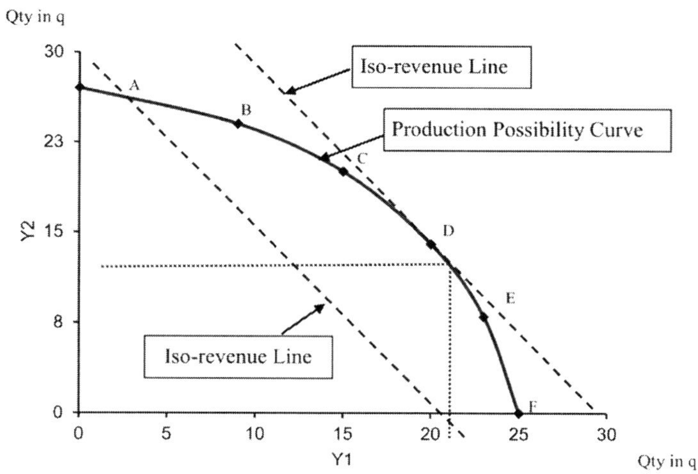

Conclusion

The Production Possibility Curve and Iso-revenue Line intersect at D Point. Thus, combination D represents **MRPS=PR**. Therefore, out of 5 acres, farmer should grow **Y_1 crop in 3 acres and Y_2 crop in 2 acres of land**.

Summary of Three Basic Production Relationships

Factor-Product Relationship	**Factor – Factor Relationship**	**Product-Product Relationship**
It is explained by Law of Diminishing Return	It is explained by Law of Factor Substitution	It is explained by Law of Product Substitution
Management Problem: **How much to produce?**	Management Problem: **How to produce?**	Management Problem: **What to produce?**
Objective : **Resource Optimization**	Objective : **Cost Minimization**	Objective : **Profit maximization**
It is concerned with Resource use efficiency	It is concerned with Resource combination and substitution	It is concerned with Resource allocation
It is used for determination of optimum input use and optimum output to produce	It is used for determination of least cost combination of resources	It is used for determination of optimum combination of products to be produced
Functional relationship: Y= f (X1 | X2, X3,…., Xn)	Functional relationship: Y= f (X1,X2,| X3,X4,., Xn)	Functional relationship: Y1= f (Y2,|X1,X2,…., Xn)
One input varied, others are kept constant	Output is constant and inputs are varied	Inputs are constant and products are varied
Three Zones of Production 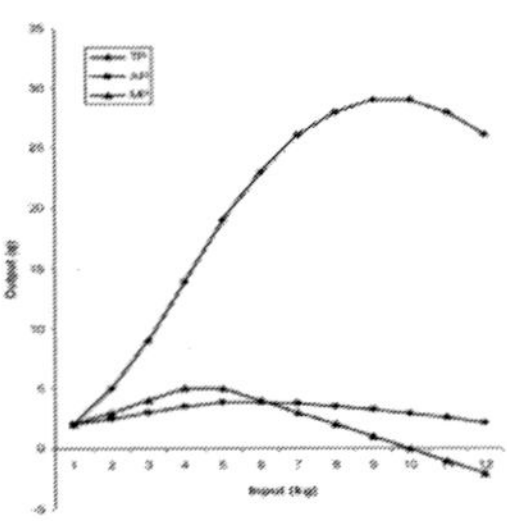	Iso-quant Curve and Iso-cost Line 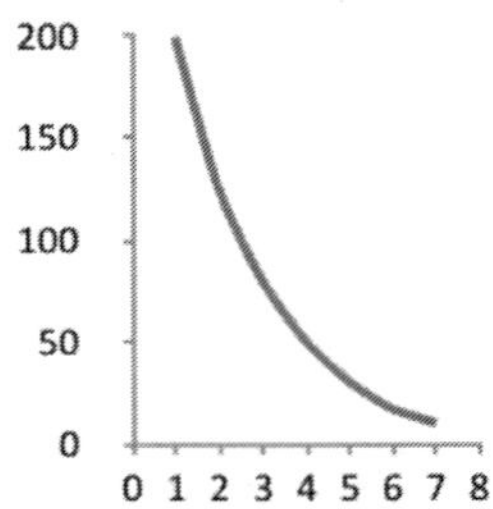	Production Possibility Curve and Iso-revenue Line 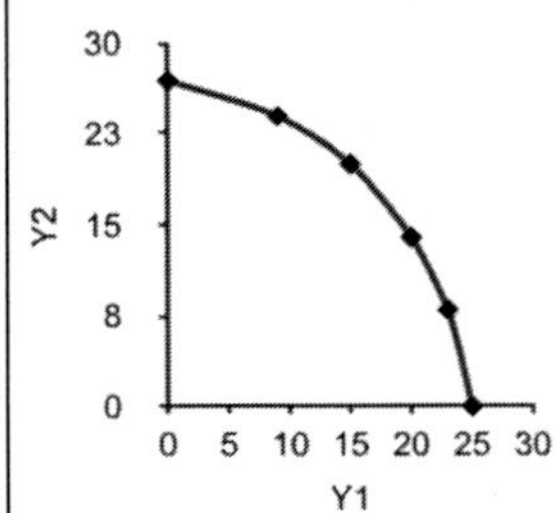
Profit rule: MC=MR	Profit rule: MRTS=PR	Profit rule: MRPS=PR

4.4. Principle of Equi-Marginal Returns

- This principle is associated with the appropriate use of **inputs** which are **limited.**
- This principle states that the **limited resources should be allocated** among alternative uses in such a way that the **marginal value product (MVP or additional returns) of the last unit of resources is equal in all uses.**
- This principle guides the farmer/manager that any limited input should be allocated in that use where it brings in the **greatest MVP,** so that the **profit can be maximized.**

Example

A farmer is having only 5 units of liquid capital each of Rs. 1000. These 5 units must be allocated among the three crops viz., bajra, maize and jowar in the follow manner based on MVPs.

Comparative Advantage (value in Rs/ha)

Amount of liquid capital used (1 unit=1000 Rs.)	Marginal value product (MVP) per unit of 1000 Rs.		
	Bajra	**Maize**	**Jowar**
1	2000 (5)	3200 (1)	2200 (4)
2	1600	2700 (2)	1800
3	1300	2300 (3)	1400
4	1100	1900	1100
5	1000	1600	800

- Allocate the inputs where it brings in the greatest MVP.
- Thus, **Rs. 3000** should be allocated to **maize** and Rs. **1000 each to jowar and bajra** to obtain maximum returns.

4.5. Opportunity Cost Principle

- **The income of next best alternative that is sacrificed** to produce any good is known as **opportunity cost**.
- Whatever must be **given up to obtain some item** is called **opportunity cost**.
- The opportunity cost is referred to as the **real cost** or **true cost** of an input.
- This principle guides the farmer/manager that If the returns from the current use of the input are **less than its opportunity cost,** then the **decision is to be changed.**

Example

(Value in Rs/ha)

Particulars	Crops		
	Bajra	**Maize**	**Jowar**
Actual cost of production	10000	11000	10000
Gross Returns	20000	23000	18000
Net Returns	10000	12000	8000
Opportunity cost	12000	10000	12000
Comparison	NR<OC	NR>OC	NR<OC
Decision	x	√	X

- Here, if the farmer is growing **bajra,** he sacrifices Rs. 12000 from maize which is the next best alternative and is called its opportunity cost. So, the actual cost is Rs. 10000, but the **real or opportunity cost is Rs. 12000**.
- Likewise, the opportunity cost of maize and jowar is Rs. 10000 and Rs. 12000, respectively.
- **If the Net Returns from the current use of the input are less than its opportunity cost, then the decision is to be changed.**

4.6 Minimum Loss Principle

- This principle is associated with the **fixed and variable cost**.
- **Fixed costs** are those costs, which do not vary with the level of output. Fixed costs incurred even in the absence of production. It includes depreciation, rental value of land, interest on fixed capital, land revenue, etc.
- **Variable costs** are those costs, which vary with the level of output. If there is no production, there are no variable costs. It includes cost of inputs, customs hiring, wages of casual labour, interest on working capital, etc.
- The farmers sometimes incur loss instead of profit **when price of farm produce goes down**. Then they should continue the farming with an objective of **minimizing losses**. Thus, this principle explains, as to how, the producer minimizes losses **under adverse price environment**
- The farmer should continue the farming till the loss is less than fixed costs. **If loss is more than fixed cost then the production should be stopped temporarily**, to minimize the loss. In the long run if selling price is less than ATC, continuous losses are incurred. In this situation, the producer should **stop the production permanently.**

Example

Costs and returns in groundnut cultivation

Price of groundnut (Rs./q)	**3000**	**2000**	**1500**
TFC (Rs/ha)	8000	8000	8000
TVC (Rs/ha)	16000	16000	16000
TC (Rs/ha)	24000	24000	24000
Production (q/ha)	10	10	10
Gross Income (Rs/ha)	30000	20000	15000
ATC (Rs./q)	2300	2300	2300
Net Profit (Rs/ha)	6000	-4000	-9000

<table>
<tr><th colspan="4">Decision whether to cultivate groundnut or not?
(on the basis of Minimum Loss Principle)</th></tr>
<tr><td>Loss if cultivated</td><td>No loss</td><td>4000</td><td>9000</td></tr>
<tr><td>Loss if don't cultivate</td><td>8000</td><td>8000</td><td>8000</td></tr>
<tr><td rowspan="3">Decision</td><td>Should be cultivated</td><td>Should be cultivated</td><td>Should not be cultivated</td></tr>
<tr><td>Because there is no loss</td><td>Because loss is less than TFC</td><td>Because loss is more than TFC</td></tr>
<tr><td>Price > ATC</td><td colspan="2">Should be permanently stopped if Price < ATC over the years</td></tr>
</table>

4.7 Principle of Comparative Advantage

The concept of comparative advantage is associated with

a. Resource productivity and

b. Cost of production of enterprise

The principle (law) of comparative advantage directs a farmer in the selection of crop and livestock enterprise in the production of which available resources have the greatest relative advantage and not absolutes advantage. Types of advantages – On the maximum net returns per hectare basis there are two types of advantages.

1. Absolute advantage.
2. Relative or comparative advantage.

Absolute advantage: Net return per hectare i.e. Total returns – Cost of production.

Comparative advantage: Refers to relative advantages of growing different crops in a region. The following table gives idea about these two advantages.

Table: Net returns of different crops in a region

S.No.	Crop Account	Bajara	Groundnut	Cotton
1.	Total income (Rs.)	3500	5000	9000
2.	Total expenses (cost)	2000	2500	3000
3.	Net returns (Rs)	1500	2500	6000
4.	Returns per rupee	1.75	2.00	3.00
5.	% returns	175%	200%	300%

From the above figures it can be said that farmers of that region can make profit by growing any of the three crops as they can get absolute advantages to the tune of Rs.1500, 2500 and 6000 from bajara, Groundnut and cotton respectively. But for making the greatest profit they will have to allocate the largest possible area under cotton alone as it has given the maximum relative returns.

Thus, if a cultivator ants to earn greatest possible returns he should produce those crops in which their relative advantage is greatest.

The specialized or diversified farming depends largely on this principle e.g. fruits & vegetables are grown largely in the vicinity of the big Cities.

5

Cost Concept and Their Interrelatioship

Abstract

This chapter examines cost concept, which refers to the sacrifice made by the producer in the form of payments, such as wages for laborers, rent for tractor and land use, and expenses for raw materials, etc. Furthermore, it explains seven cost concepts and CACP cost concept with graphical presentation

5.1. Introduction

A functional cost analysis is a method that can be applied to examine the component costs of a product in relation to the value as perceived by the customer. The outcome of the analysis is to improve the value of the product while maintaining costs and/or reduce the costs of the product without reducing value. The functional cost analysis is a value engineering method that aims to increase the difference between the cost and the value of a product. The cost is the amount that is incurred in the production and delivery of the product. This expense can include the price of parts, labour, overhead (e.g., building, power), packaging, shipping, and advertising, among others. What the product is worth in the eyes of the customer is considered the value.

5.2. Meaning of Cost

A producer has to work very hard to produce goods service. He has to make a lot of effort in the process. In the beginning, the producer must arrange money to organize the production activity. We have already said that various factors in the form of land, labour, and capital are required to produce goods. These factors are not available for free. The producer must purchase them in the right quantity needed for production. Similarly, raw materials and other things have to be purchased and stored. To purchase these things, the producer must be ready to pay the price for them. The term 'cost of production' means expenses incurred in the production of a commodity. Thus, the cost of production of a commodity is the aggregate of prices paid for the factors of production used in producing a commodity. This refers to the total amount of money spent on the production of a commodity. The determinants of cost of production are: the size of plant, the level of production, the nature of technology used, the quantity of inputs used, managerial and labour efficiency. Thus, the cost of production of a commodity is the aggregate of prices paid for the factors of production used in producing a commodity.

5.3. Definition of Cost

Cost is defined as the money expenditure incurred by the producer to purchase (or hire) factors of production and raw materials to produce goods and services. In a way, cost is a kind of sacrifice made by the producer and the sacrifice was made in terms of making payments; such as wages to labourers, rent for the use of land and tractor and incurring expenditure on raw materials etc.

5.4. Cost Function

A cost function is a mathematical formula used to use to chart how production expenses will change at different output levels. In other words, it estimates the total cost of production given a specific quantity produced. The relationship between output and costs is expressed in terms of cost function. By incorporating prices of inputs into the production function, one obtains the cost function since cost function is derived from production function. However, the nature of cost function depends on the time horizon. In microeconomic theory, we deal with short run and long run time. According to Prof. Watson, the cost function expresses a functional relationship between costs and output that determine it. Symbolically, the cost function is

$C = f(Y)$

Where, C = Cost and Y = Output

5.5. Cost of Production

Costs of Production are total sum of money required for the production of a specific quantity of output. The expenditure incurred in producing a unit quantity of output is known as cost of production, for example, Rs. xx /kg of Rs. xx /quintal.

Following elements are included in the cost of production;

- Rent of land.
- Wages of labour.
- Interest on capital.
- Wear and tear of the machinery and building.
- Advertisement charges.

- Insurance charges.
- Payment of taxes.

5.6. Features of Cost

Following are the main features of cost

1. **Explicit Costs and Implicit Costs:** The costs of production from the point of view of individual firm are divided into two parts;

a. **Explicit Costs:** Explicit cost represents all such expenditure which are incurred by an entrepreneur to pay for the hired services of factors of production and in buying goods and services directly. The explicit costs include wages and salary payments, expenses on the purchase of raw materials, light, fuel, advertisements, transportation, taxes and depreciation charges. In short, all the items of expenses appearing on the debit side of trading account of a firm represent explicit costs. Explicit cost is also called accounting costs.

b. **Implicit Costs:** The implicit costs are imputed value of the entrepreneur's own resources and services. Implicit costs can be defined as expenses that an entrepreneur does not have to pay out of his own pocket but are costs to the firm because they represent an opportunity cost. Implicit costs thus are the alternative costs of the self-owned and self-employed resources of a firm.

2. **Real Cost:** Real cost is a subjective concept. It expresses the pains and sacrifices involved in producing a commodity. The money paid for securing the factors of productions is money cost whereas the efforts and sacrifice made by the capitalists to save and invest, by the workers in foregoing leisure and by the landlords constitute real cost.

3. **Sunk Costs:** A sunk cost is a cost that has already occurred and cannot be recovered by any means. Sunk costs are independent of any event and should not be considered when making investment or project decisions. Sunk costs cannot be recovered if a business decides to leave an industry. All sunk costs are considered fixed costs. However, it is important to realize that not all fixed costs are considered sunk costs. Recall that sunk costs cannot be recovered. Take for example equipment (a fixed cost). Equipment can be resold or returned at a determined price; therefore, it is not a sunk cost. Sunk cost is also known as past cost, embedded cost, prior year cost, stranded cost, sunk capital, or retrospective cost. Examples of Sunk Costs: Marketing study, Research and development, Training and Hiring bonus.

4. **External Costs:** This is the cost which is imposed on a third party. For

example, if you smoke, some people may suffer from passive smoking. That is the external cost.

5. **Private Costs:** The costs you pay. e.g. the private cost of a packet of cigarettes.
6. **Social Costs:** This is the total cost to society. It will include the private costs plus also the external cost (cost incurred by a third party). May also be referred to as 'True costs'.
7. **Social Marginal Cost:** The total cost to society of producing one extra unit. Social Marginal Cost (SMC) = Private marginal cost (PMC) + External marginal Cost (XMC)
8. **Economic Cost:** The economic cost includes not only the explicit cost but also the implicit cost. The money rewards for the own services of the entrepreneur and the factors owned by himself and employed in production are known as implicit costs or imputed costs. The normal return on money capital invested by the entrepreneur, the wages or salary for his own services and rent of the land and buildings belonging to him and used in production constitute implicit cost.

   Thus, Economic cost = Explicit cost + Implicit cost.

a. It may be pointed out that the firm will earn economic profits only if it is making revenue in excess of economic cost.
b. **Economic Profit** = Total Revenue - Economic Costs

9. **Opportunity Cost:** The opportunity cost of any goods is the next best alternative goods that is sacrificed. The loss of other alternatives when one alternative is chosen.

   Opportunity Cost = (Return of next best alternative not chosen) - (Return of option chosen)

   One relative formula for the calculation of opportunity cost could be

$$\text{Opportunity costs} = \frac{\text{(What you sacrifice)}}{\text{(What you gain)}}$$

   If we think about the cost of opportunity like this, then the equation is very easy to understand and it's straight forward.

   For example : you may have the choice between two jobs, a mechanicor bartender. These figures are purely hypothetical and serve for the purpose of illustration of how to calculate opportunity costs. If you work as a mechanic, you could earn Rs.50 per hour, if you are a bartender, even with tips, your wage could be around Rs.

25 per hour. At first, it may look like Rs. 25 per hour. But this is not the way opportunity costs are calculated. The formula is not "what I sacrifice minus what I gain." Instead, it is necessary to look at the ratio of sacrifice to gain. Going back to our example, if you chose to spend an hour working as a bartender instead of a mechanic, then you are actually giving up (Rs. 50 mechanic / Rs. 25 bartender) = Rs. 2 of opportunity cost. This Rs. 2 says, for every year, I earn working for one hour as a bartender; I sacrifice Rs. 2 working the same hour as a mechanic.

10. **Replacement Costs: -** Replacement cost is also known as depreciation cost. With the continue utilization of fixed assets such as tools, equipments and machinery, they depreciated. Hence, there is need to replace these assets. The cost at which these assets are replaced is known as replacement cost.
11. **Production Costs:** Production costs are the costs which are incurred to increase the level of output of a firm. The payments which are made to factors of production are known as product costs.
12. **Selling Costs:** Selling costs are incurredto increase the sale of available products.Examples are commission paid to salesman, advertisement of different products.
13. **Fixed Cost and Variable Cost:** Fixed and variable factors are with reference to short run production function. Short run is a period of time over which certain factors of production cannot be changed, and such factors are called fixed factors. The costs incurred on fixed factors are called fixed costs. The factors whose quantity can be changed in the short run are variable factors, and the costs incurred on variable factors are called variable costs. Fixed costs are those which are independent of output, that is, they do not change with changes in output. These costs are a 'fixed' amount, which must be incurred by a firm in the short run whether the output is small or large. E.g. contractual rent, interest on capital invested, salaries to the permanent staff, insurance premium and certain taxes. Variable costs are those costs, which are incurred on the employment of variable factors of production whose amount can be altered in the short run.

    Thus, the total variable costs change with the level of output. It rises when output expands and falls when output contracts. When output is nil, variable cost becomes zero. These costs include payments such as wages of labour employed, prices of raw materials, fuel and power used and the transport costs.

5.7. Seven Cost Concept

1. Fixed Costs

Fixed costs are those costs that do not change in the short run period of time. Fixed costs remain the same regardless of the amount of production and sale of commodities. These costs are incurred by the company irrespective of its production, i.e. even at zero production, the firm incurs fixed cost.

A fixed cost can be defined as follows: "A fixed cost is the cost that remains the same and fixed irrespective of the production of goods."

These are also known as indirect costs, sunk costs and overhead costs. The summation of all these costs is called total fixed costs (TFC). TFC is a horizontal straight line parallel to X-axis.

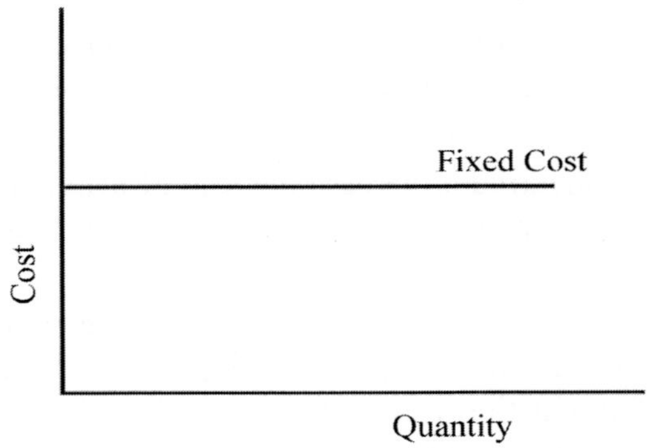

Fig. 1: Fixed Cost

Uses of fixed cost

1. Useful in evaluating break – even analysis;
2. Helps in pricing strategy;
3. Helps in decision – making;
4. Helps in controlling variable costs;

2. Variable Costs

A variable cost is that cost which changes in short – run and long – run time period. It always keeps on changing. These costs are incurred during production process and thus are the costs incurred for employing various factors of production. A fixed cost becomes a variable cost in the long – run.

A variable cost is defined as follows: - "A variable cost is the expenditure incurred on the production of goods and therefore is ever changing."

These are also known as working costs, operating costs, direct costs, prime costs, circulating costs and running costs. These are second phase costs. The summation of these costs refers to total variable costs (TVC). Graphically TVC has inverse 'S' shape (Figure 2).

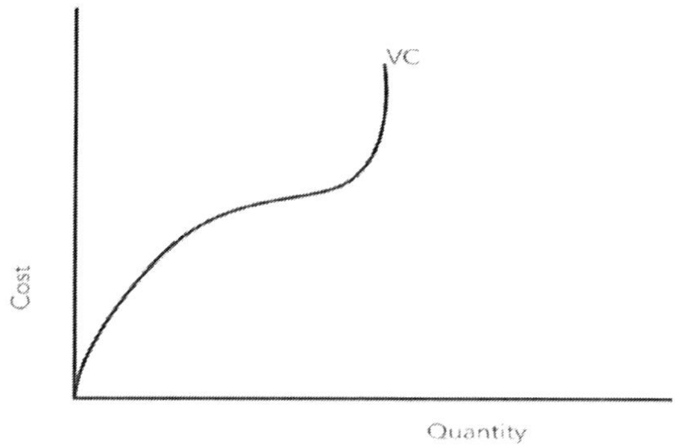

Fig. 2: Variable Cost

Uses of variable cost:

1. Helps to set prices for the commodity;
2. Helps to plan profits;
3. Helps in decision making;
4. Helps in cost control;

Difference between fixed cost and variable cost:

Fixed cost	**Variable cost**
1. **Meaning:** Fixed costs are costs that arefixed and have to be incurred irrespectiveof production of goods.	Variable costs are costs that are not fixed and keep on changing as per the production of goods.
2. **Nature:** These costs are fixed.	These costs are not fixed.
3. **Occurrence:** These costs occur even when the there is no production of goods.	These costs occur only when there is production of goods.
4. **Changes:** Fixed cost only in the long run.	Variable cost changes in short run as well as long run.
5. **Other names:** Other names for fixed costs are Overhead costs or supplementary costs.	Other names for variable costs are Prime costs.
6. **Examples:** Depreciation charges, maintenance costs, property taxes, interest, rent, etc.	Electricity charges, raw material purchases, conveyance, salaries and wages, etc.

3. Total Costs

Total cost is the total expenditure incurred by the producer to produce his goods. Total cost is also the summation of total fixed costs and total variable costs. Total cost is evaluated as follows:

1. Total Cost = Cost per unit x Quantity Produced
2. Total Cost = Total Fixed Cost (TFC) + Total Variable Cost (TVC)

Total cost is defined as follows:

"Total cost is the cost which is incurred by the producer to produce a particular quantity of the commodity." Its shape is similar to that of TVC (Figure 3).

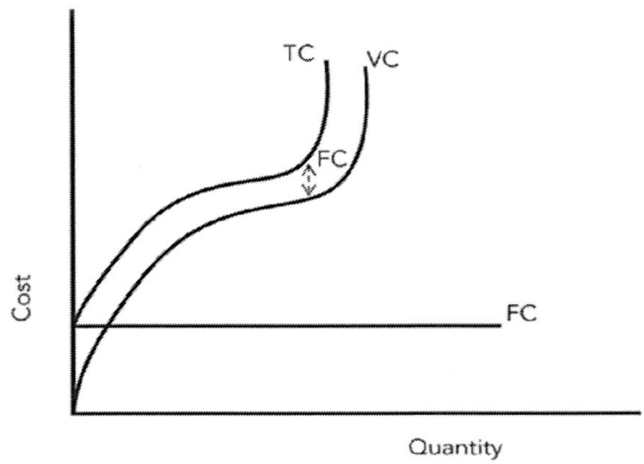

Fig. 3: Total Cost

Uses of total cost:

1. Helps in economies of scale as a producer needs large number of raw materials for large production;
2. Helps in pricing policy;
3. Helps in decision – making;

4. Average Variable Cost (AVC)

It is the amount spent on the variable inputs to produce an unit of output. Algebraically it is expressed as :

$$AVC = \frac{\text{(Total variable cost)}}{\text{Output}} = \frac{TVC}{Q}$$

When a small amount of output is produced, cost of variable input per unit of output becomes very high. This is to say in other words, that productivity of variable input increases when greater amounts are used in the production of the commodities due to economies of scale. This causes AVC to have 'U' shape when it is graphed. This is shown in the Figure 4. When it is 'U' shaped it becomes reciprocal of average physical product curve.

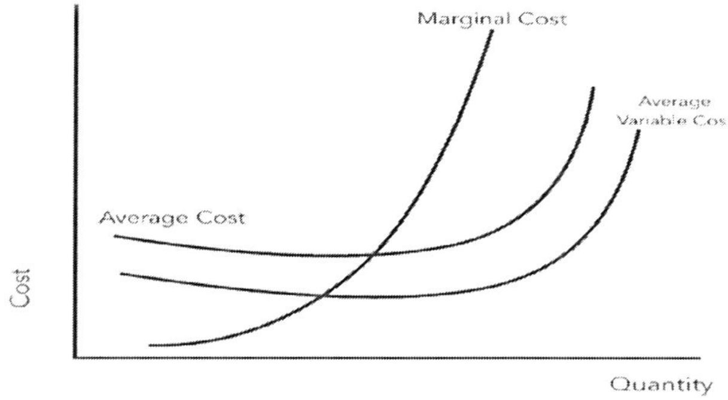

Fig. 4: Average Variable Cost

AVC falls to minimum level at the output level where APP is maximum. There. after due to production of greater amount of output, AVC rises again and becomes vertical at certain level of maximum output.

5. Average Fixed Cost (AFC)

It is the cost of fixed resources or inputs required for producing one unit of output and it is given by the formula as

$$AFC = \frac{\text{(Total fixed cost)}}{\text{Output}} = \frac{TFC}{Q}$$

AFC curve is declining with the increased output because TFC is constant. Due to this it is continuously falling up to its maximum output. It is having the shape of hyperbola (Figure 5).

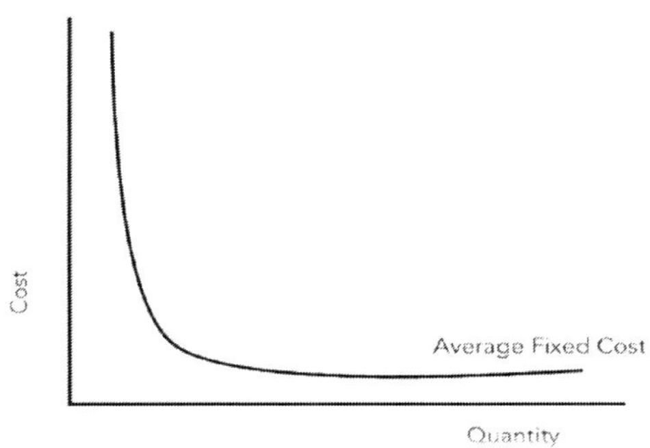

Fig. 5: Average Fixed Cost

6. Average Total Cost or Average Cost (ATC or AC)

When the total costs are divided by output, we get ATC (Figure 6).

$$ATC = \frac{\text{Total Cost}}{\text{Output}} = \frac{TC}{Q} = \frac{TFC+TVC}{Q}$$

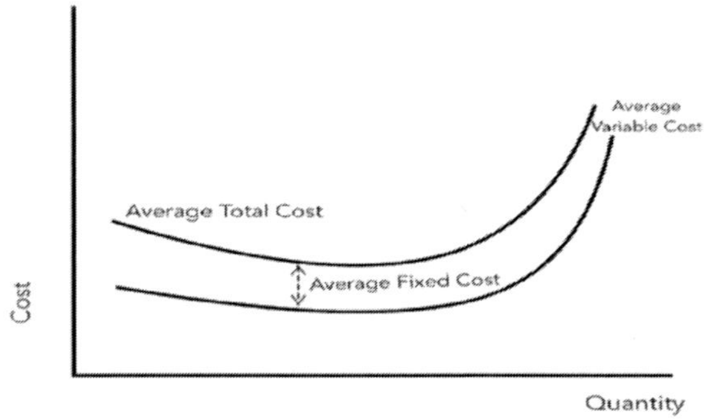

Fig. 6: Average Total Cost

7. Marginal Cost (MC)

Marginal cost is the expenditure incurred by the producer to produce an additional or an extra unit of the commodity. Marginal cost is the additional cost incurred for producing one extra unit after producing certain amount of units.

MCn = TCn – TCn-1

Marginal cost is defined as follows: "Marginal cost is the cost or expense incurred for producing an additional or an extra unit of a commodity." Algebraically it is expressed as

$$\frac{\text{Change in total cost}}{\text{Change in output}} = \frac{\Delta TC}{\Delta Q} = \frac{\Delta TVC}{\Delta Q}$$

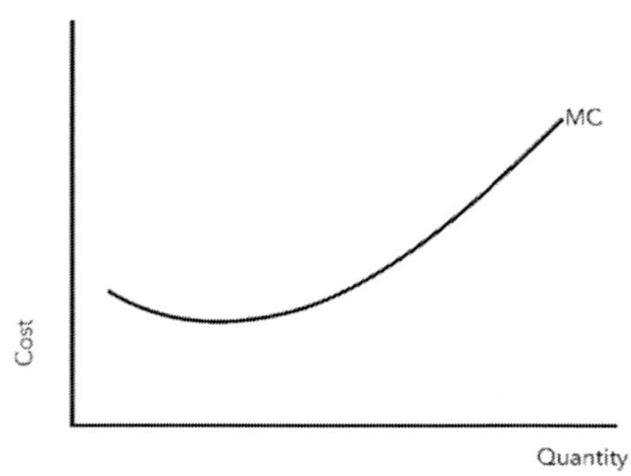

Fig. 7: Marginal Cost

Uses of marginal cost

1. Helps in decision making
2. Helps to determine costs for each commodity
3. Helps in planning profits.

Difference between total cost, average cost and marginal cost:

Total cost	Average cost	Marginal cost
1. **Meaning:** Total cost is expenditure incurred to produce a bulk quantity of the commodity.	Average cost is the expenditure incurred to produce one unit of the commodity.	Marginal cost is expenditure incurred to produce an extra or additional unit of the commodity.
2. **Definition:** Total cost is the cost which is incurred by the producer to produce a particular quantity of the commodity	Average cost is the expense incurred by the producer to produce one unit of the total production.	Marginal cost is the costor expenseincurred for producing an additional or an extra unit of a commodity.
3. **Includes:** It includes total fixed costs and total variable costs.	It includes average fixed costs and average variable costs.	It includes only the cost for producing an extra unit.
4.**Calculation:** It is calculated in two ways.	It is calculated in two ways.	It is calculated in one way.
5. **Formulae:** TC = TFC + TVC	AC = AFC + AVC	MCn = TCn – TCn-1
6. Graphs: 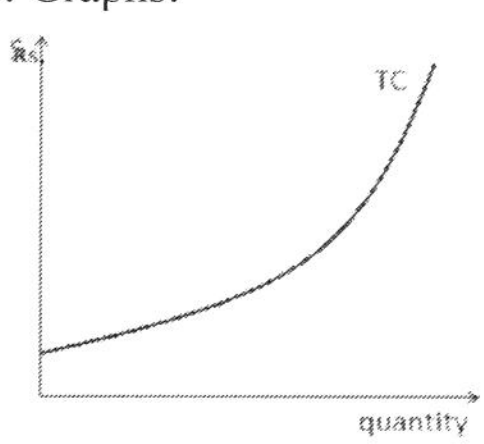	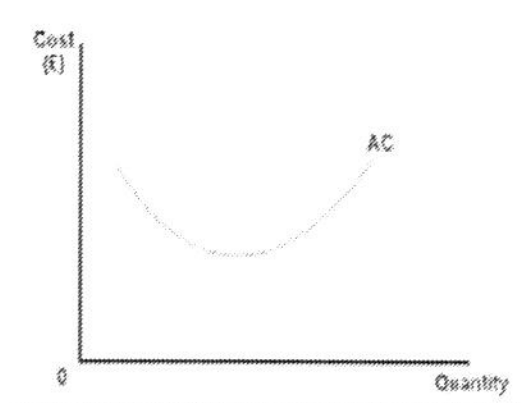	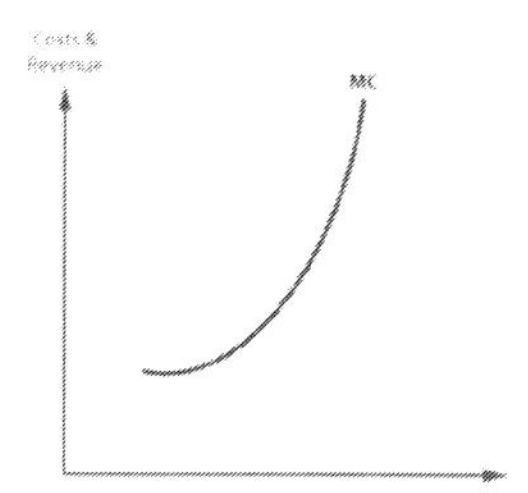
7.**Uses:** • Helps in economies of scale as a producer needs large amount of raw materials for large production; • Helps in pricing policy; • Helps in decision - making;	• Helps to cut down excess expenditure, as per unit cost is calculated; • Helps in optimum utilization of resources; • Helps in pricing strategy.	• Helps in decision making; • Helps to determine costs for each commodity; • Helps in planning profits.

5.8. Cacp Cost Concept

Cost concepts are widely used because of their relevance in the decision-making process. This means that these costs serve as a basis to expand the size of the farm, to buy the requisite capital assets in the long run and the requisite inputs in the short run. For example, variable costs have a bearing on the level of production in the short run, on the other hand the decisions like expanding the size of the farm, buying the durable assets, etc., are based on the total costs. Crop costs or costs related to livestock production are split up into various cost components, such as cost A_1, A_2, B_1, B_2, C_1, and C_2. Before knowing the cost concepts, let us know the different costs involved in the production of crop and livestock products.

Labour Costs: These include wages paid in cash and kind, both to casual, contract and permanent labourers. These costs come under paid out costs. Farmers do not pay for their own labour, such as family labour, owned bullock labour, owned machine labour, etc. The wages for these family labourers are imputed based on market wage rate. Labour cost is a variable cost.

Machinery Costs: These include hiring charges for machinery, implements, etc. Fuel and lubricant charges of the machines are considered as variable costs, whereas repairs and maintenance of machines, insurance premium, taxes, license, depreciation, and interest on capital invested are considered as fixed costs. For owned machinery, the existing market hiring charges should be used as variable costs.

Livestock Costs: These include cost of veterinary medicine, cost of fodders and feeds, labour services used, etc., in the maintenance of the livestock. These are generally considered as variable costs. In contrast the interest rate paid on the amount spent to purchase the animals and depreciation on the animal are considered as fixed costs.

Land Costs: These refer to rental value of owned land, and rent of the lease-in-land.Rent of the owned land is imputed hence it is called rental value of owned land, but not the rent. Rent of the land arises only when the land is leased in. Rent of the land* leased out is not land cost, but it is considered as income to the farmers. Rental value of owned land is considered as a cost though it is not incurred by the farmer, because, we use the concept of opportunity cost and impute the existing rent of similar farms in the region.

Building Costs: These include interest on the value of the buildings and the depreciation on the buildings. Both these costs are regarded as fixed costs.

Input Costs: Both farm produced and purchased inputs viz., seeds, manures, plant protection chemicals, weedicides, etc., are estimated for their current value at market prices.

Interest on Working Capital and Fixed Capital: By definition working capital vary with the level of production while fixed capital remains the same regardless of the level of production. These respective interest rates should not be clubbed into one as these are entered separately while deriving the cost concepts. Normally in farming the inputs are not used at a time but at different points of time according to the requirements of the crop. Under such circumstances, the interest calculated on the working capital should be reduced for half of the crop period. It is more an approximate method than a realistic method.

Cost Concepts: These include cost A_1, A_2, B_1, B_2, C_1, and C_2.

Cost A_1: The following items are included in cost A1

1. Wages of hired human labour.
2. Wages of permanent labour.
3. Wages of contract labour.
4. Wages of hired bullock labour.
5. Imputed value of owned bullock labour.
6. Charges of hired machinery.
7. Imputed value of owned machinery.
8. Market rate of manures and fertilizers.
9. Market rate of seeds.
10. Imputed value of owned seeds.
11. Imputed value of manures.
12. Market value of pesticides, herbicides, hormones, etc.
13. Irrigation charges.
14. Land revenue, cess and other tax.
15. Depreciation on farm machinery, implements, equipment farm buildings.
16. Irrigation structures, etc.
17. Interest on working capital.
18. Other miscellaneous expenses.

The following concepts can be used for easy calculation of the cost of cultivation:

1. Depreciation for buildings: 2 per cent for pucca building; 5 per cent for tiled building and 10 per cent for kuccha building.

2. Depreciation for implements: 10 per cent for major implements and 20 per cent for minor implements.
3. Depreciation for cattle: Appreciation in the value of animals during the first 3 years would be at the ratio of 1:3:5. It remains constant during 4th and 5th year. Then it is assumed that the value of animal depreciates @ 12.5 per cent per year from 6th to 14th year in straight line method.
4. Interest on working capital: 12 per cent per annum or opportunity cost of capital.

Cost A_2 = Cost A_1 + Rent paid for leased in land.

Cost B_1 = Cost A_1 + Interest on the value of owned capital assets (excluding land). Interest rate of long - term government floated loans or securities: 10 per cent.

Cost B_2 = Cost B_1 + Rental value of owned land (less land revenue) and Rent paid for leased in land.

Cost C_1 = Cost B_1 + Imputed value of family labour.

Cost C_2 = Cost B_2 + Imputed value of family labour.

Cost C_3 = Cost C_2 + 10 percent of cost C_2 to account for managerial cost of inputs of farmers.

Income measures in relation to different cost concepts:

1. Farm Business Income = Gross Return - Cost A_1.
2. Owned Farm Business Income = Gross Return - Cost A_2.
3. Family Labour Income = Gross Return - Cost B_2.
4. Net Income = Gross Return - Cost C_2.
5. Farm Investment Income = Net Income + Imputed rental value of owned land + Interest on fixed capital.

6

Farm Business Analysis

Abstract

This chapter explores how to measure and enhance farm income and profitability in the agricultural and livestock sector. It applies technical and financial efficiency indicators to evaluate the performance of different farm enterprises. It also explains how to maintain farm inventory and records

6.1. Meaning and Concept of Farm Income and Profitability

Farm income is the money earned from an agriculture business that typically gets reported separately from other types of income for tax purposes. Many different types of farming businesses exist, such as those that raise livestock, cultivate vegetables, or grow Christmas trees.

A farm income statement (sometimes called a farm profit and loss statement) is a summary of income and expenses that occurred during a specified accounting period. This period is usually the calendar year for farmers (January 1 - December 31). Some farms are eligible for special farm tax credits and other tax breaks.

- Farm income refers to the money generated by farm or agribusiness operations.
- Farm income is treated a bit differently than non-farm income for tax purposes.
- Farmers are required to fill out a Schedule F on their tax returns to report farm income.

Farm income refers to profits and losses that are incurred through the operation of a farm or agricultural business. In addition to direct revenue such as selling crops, farm income can include earnings from other sources, such as crop insurance.

6.2. Profitability

Three main indicators of profitability are shown directly in the income statement and these are: Net farm income, off – farm income and Net Income.

i. Net Farm Income = Total farm Revenue – Total farm cost Net Farm Income refers to the 'Bottom line' profit that is earned (or projected to earn) by the business during the accounting period.

ii. Off-farm income comes from non-farm jobs and custom work on other farms

iii. Net income is net farm income plus off –farm income minus all income taxes and social security payments due on these income sources.

Net income shows what is available to pay for principal on past debts, new farm investments, family off-farm investments such as retirement funds, stocks, mutual funds etc, Net income is the real bottom line of a business and is the single most important indicator of profitability.

Ratio indicators

The remaining measures of profitability are all ratios. These are return on assets, return on equity, growth in networth and the profit margin ratio

1. Return on Assets (ROA)

It is a measure of profitability, measuring the rate of return that the farm business earns on its average asset over the period, the higher the return, the more profitable the farm business.

$$\text{Return on Asset} = \frac{\text{Net farm income + interest expense - labour}}{\text{Total farm assets (average)}} \times 100$$

2. Return on Equity (ROE)

It is a measure of the return to the networth (equity) in the business. The farm equity is the capital that could be invested elsewhere (if you are not farming) and so this analysis provides an interesting perspective to see just how good a return you are receiving on your investment in farming as compared to other alternatives.

$$\text{Return on equity} = \frac{\text{Net farm income - labour}}{\text{Farm equity (average)}} \times 100$$

3. Expense /Return Ratio

Shows the percentage of the farm income that is required to cover the operating expenses, excluding the principal and interest payments.

$$\text{Expenses Return Ratio} = \frac{\text{Operating Expenses}}{\text{Value of farm production}} \times 100$$

The value of farm production is the total value of farm sales less the cost of purchased feeds, grain and market livestock.

All these profitability measurements provide information on how well the business is doing, but they do not show whether the resources are used efficiently. This is where the final set of indicators comes in.

6.3. Technical and Economic Efficiency Measures in Crop and Livestock Enterprises

Farmer makes several decisions in the use of resources. Mere decision-making may not result in profitable organization of the farm business. There is need to evaluate the decisions of the farmer for which, we need some guides and standards. Here comes the role of efficiency measures, which are categorized into technical efficiency measures and economic efficiency measures. These are explained here under.

6.3.1. Technical Efficiency Measures

Broadly, three types of technical inputs wiz., land, labour and capital are used on the farms. Every farmer aims at the efficient utilization of these inputs. To measure the efficiency, we need some criteria or yardsticks or indicators. Following are the various measures to express the efficiency of these technical inputs.

1. Land

Generally, land is measured in terms of acres or hectares owned by the farmer. Always the entire land is not put to cultivation due to the inadequacy of resources. When part of the land is slack, the efficiency of the farm is affected. Land use efficiency is measured by productivity, cropping intensity and crop yield index.

(i) ***Productivity of Land***

The production efficiency with respect to a particular crop, is expressed as the ratio between actual yield of a crop on the farm and average yield of the same crop in the locality. This relationship is expressed in proportionate terms.

Example:

Paddy yield/ ha. on a farm = 40q

Average yield of paddy /ha in the locality = 35q

$$\text{Production Efficiency} = \frac{\text{Actual crop yield on the farm}}{\text{Average yield in the locality}} \text{x}100$$

$$= \frac{40}{35} \; 100 = 114.28\%$$

(ii) *Cropping Intensity*

It measures the extent of land use for the production of the crops during an agricultural year. Cropping intensity is worked out from the following equation

$$\text{Cropping Intensity} = \frac{\text{Gross cropped area during the agricultural year}}{\text{Net cropped area}} \text{x}100$$

For example, if a farmer is raising three crops in an agricultural year in his 10 hectare of irrigated land, then the cropping intensity is

$$\frac{30}{10} \times 100 = 300\%$$

(iii) *Crop Yield Index*

It is a good measure of index to compare the yields of all the crops grown on the farm with that of average yields of the same crops grown in that locality. The index is measured in terms of percentage. The computation of crop yield index is illustrated in Table 1.

Table 1: Estimation of Crop Yield Index.

Crop	**Average yield**		**Crop acreage on selected farm (ha.)**	**Production efficiency**	**Production efficiency × crop acreage**
	Given locality	**Selected farm**			
Paddy (Q)	68	80	4	117.65	470.60
Sugarcane (tonnes)	90	110	6	112.22	733.32
Total	-	-	10	-	1203.92

Crop yield index = 1,203.92/10 = 120.39.

Thus, the farm in question is more efficiently organized in terms of crop yields as compared to an average farm in the area since the crop yield index is more than 100.

2. Labour

The profitability of the farm business is influenced by the efficiency of the labour resource. Farmers who do not introduce new methods to achieve higher labour efficiency find it difficult to continue in the agricultural profession as the wages of farm labour are on the increasing trend. Crop acreage per man equivalent year, livestock maintained per man, gross income per man and productive man work units per man equivalent year are the important measures of labour efficiency.

Farmers use different kinds of labour for executing various farm operations. These include men labour, women labour, and child labour. However, labour is measured in terms of man-days. A man-day is the productive work accomplished by a worker in eight hours of a day assuming average efficiency. As we are employing not only men labour but also women and child labour on the farm, women and children should be converted into man-days, which is a standard unit of labour measurement. Taking the efficiency into consideration, three women are considered as equivalent to two men and two children are equal to one man. All woman-days and child-days should be converted into man equivalent days by taking above ratio into consideration. The sum of all man-days and man equivalent days is divided by 365 days to get man equivalent year.

For example, paddy crop requires 100 men labour, 120 women labour and 50 child labour per hectare.

Man days	=	100
Woman labour converted into man equivalent days	=	120×2/3
	=	80 man equivalent days
Child labour converted into man equivalent days	=	50×1/2
	=	25 man equivalent days
Sum of man days + man equivalent days	=	100 + 80 + 25 = 205
Man equivalent year (or) person year = 205/365	=	0.56
Man equivalent year or person year can also be computed in the following manner		
◦ Farmer's own contribution of labour	=	4 months
◦ Family labour contribution	=	8 months
◦ Hired labour contribution	=	12 months
◦ Total contribution = 4 + 8 + 12	=	24 months
◦ Man equivalent year or person year	=	$\frac{\text{Total labour contribution}}{\text{12 months}}$
	=	$\frac{24}{12} = 2$

Labour Efficiency Measures

i. ***Crop Acreage per Man Equivalent Year:*** It is one of the simplest methods of measuring the productivity of labour input and is calculated by dividing the number of acres under crops by man equivalent year. Higher the ratio greater the productivity or efficiency.

ii. ***Livestock Maintained per Man:*** It is arrived at by dividing the total livestock maintained on the farms divided by man equivalent year.

iii. ***Gross Income per Man:*** It is calculated by dividing gross incomobtained on thefarm by man equivalent year.

iv. ***Productive Man Work Units per Man Equivalent Year (PMWU):*** Assuming average efficiency, the work done on the farm by the worker in a day of eight hours is known as productive man work unit i.e., man-day.

To work out total productive man work units, we require information on the average requirement of productive man work units per unit for all the enterprises taken up on the farm, as well as the size of each enterprise. Example is presented in Table 2.

Table 2. Estimation PMWU Per man Equivalent Year

Name of enterprise	Area (in ha)	Average requirement of productive man works units per unit	Total productive man work units (man days × size of the enterprise)
Paddy	5	250	1,250
Groundnut	4	120	480
Dairy	3	40	120
Total	-	-	1,850

$$\text{PMWU per man equivalent} = \frac{\text{Total productive man works units}}{\text{Man equivalent year}}$$

6.3.2. Economic Efficiency Measures

Though the technical efficiency in the use of resources is essential, but the economic efficiency ultimately decides the success or failure in the organization of the farm business. The economic efficiency measures are classified into aggregate and ratio measures.

1. Aggregate Measures

Aggregate measures are the measures meant to reflect the income earned by the various factors of production for the farm as a whole. Here the entire farm is considered as a unit, and efficiency is worked out for the entire farm in terms of each resource used. Following are the aggregate measures:

i. ***Total Capital Invested:*** Investment on fixed assets (land, building, etc.) and working assets (machinery, livestock, equipment, etc.) together represent the size of the farm.

ii. ***Total Value of Production (gross income):*** It is the total income derived from the sale of main product as well as by-products from the enterprises taken up by the farmer in an year. This includes value of home consumed products plus the value of products sold. This is a better measure of comparing the different farms. It is computed as shown in Table 3.

Table 3: Computation of Total Value of Production

Crop	Area (ha)	Main Product	By-product (CL)	Price per unit (Rs.)		Gross income (Rs.)
				Main Product	By-product	
1	2	3	4	5	6	7
Paddy	1.0	35 qtls	4	700	250	25,500
Sugarcane	2.0	90 tonnes	-	900	-	81,000
Total						1,06,500

iii. ***Total Expenses.*** These include working costs and fixed costs of the farm. In general, fixed costs are worked out for the entire farm and working costs are estimated crop-wise in terms of inputs, such as, labour, seeds, fertilizers, etc., and totalled up for the whole farm. Fixed costs include deprecation, land revenue, interest on fixed capital, rental value of owned land, insurance, etc.

iv. ***Net Worth:*** Net worth is the surplus of total assets over the total liabilities.

Net worth or equity = Total assets - total liabilities.

The high level of net worth indicates strong financial position of the farm at a point of time. Balance sheets prepared over the years indicate financial progress of a farm business.

v. ***Net Cash Income:*** The surplus of cash income over cash operating expenses indicate net cash income. This is the amount available with the farmer for future investment on the farm. Net cash income = Gross cash income-cash expenses.

vi. ***Net Farm Income:*** It is worked out as Net cash income ± change in the inventory-depreciation.

vii. ***Farm Earnings:*** It is computed as: Net farm income + value of farm products consumed at home.

viii. ***Family Labour Earnings:*** It is farm earnings less interest charges on fixed capital.

ix. ***Returns to Management:*** It is computed using the following equation.

Returns to management = Family labour earnings - imputed value of family labour.

2. Ratio Measures

These measures reflect per unit returns or costs to the inputs involved in the organization of the farm business.

i. **Fertilizer Costs per Crop Acre:** It is the ratio of total expenditure on fertilizers to number of acres under crop. This varies from crop to crop and farm to farm. In reality, farmers do not have exact nutrient status of the soil but however ambitious farmers apply the fertilizers in more than required level. Under such situation, efficiency of fertilizer is brought down leading to the increase in the cost of fertilizer.

ii. **Power and Equipment Cost/Crop Acre:** This relates to running costs incurred in operating power drawn machinery per acre of cropped area. The running costs vary with make, condition and age of the machine. Similarly, some commercial crops require intensive use of machinery. Hence comparison becomes relevant only when similar farms using the same implements for the same crops are found.

iii. **Cost Ratios:** These ratios provide information to the manager to decide whether the existing costs of the business farm are higher or lower. These ratios indicate what proportion of gross income is spent for meeting different types of expenditure. These ratios are operating ratio, fixed ratio and gross ratio. The following expressions are used to compute them.

a. $$\text{Operating ratio} = \frac{\text{Operating expeness on the farm}}{\text{Gross income on the farm}}$$

b. Fixed Ratio: It gives the proportion of fixed expenses in gross income of the farm. This ratio is worked out as:

$$\frac{\text{Fixed costs on the farm}}{\text{Gross income on the farm}}$$

c. Gross Ratio: Gross ratio is the ratio of total costs of the farm to the gross income. The relevant formula to work out gross ratio is:

$$\frac{\text{Total cost on the farm}}{\text{Gross income on the farm}}$$

iv. **Solvency Ratios:** These ratios measure the degree of financial safety (solvency) and liquidity of comparable business farms at a point of time. These ratios include current ratio, working ratio, net capital ratio and debt-equity ratio.

a. $$\text{Current ratio} = \frac{\text{Current assets}}{\text{Current liabilities}}$$

b. $$\text{Working ratio} = \frac{\text{Current assets + intermediate assets}}{\text{Current liabilities + intermediate liabilities}}$$

c . $$\text{Net capital ratio} = \frac{\text{Total assets}}{\text{Total liabilities}}$$

d. $$\text{Debt-equity ratio} = \frac{\text{Total liabilities}}{\text{Owners equity}}$$

v. **Income Ratios**

Net Income per Acre: Another important income ratio is the net income per acre. To compute this ratio we require information on (1) gross income of the farm (2) total cost on the farm; and (3) total acreage of the farm. Net income per acre is gross income less total costs. Now the net income per acre is worked out as :

$$\frac{\text{Net income}}{\text{No.of acres}}$$

Higher level of this ratio indicates greater efficiency of the farm.

vi. **Capital Ratios**

a. ***Capital per Unit of Gross Income:*** It is computed by dividing total capital invested with gross income.

b. ***Capital per Man:*** It is worked out by dividing the total capital invested by the number of man equivalent years.

c. ***Rate of Capital Turnover:*** It indicates the number of years it will take for the farm to return the investment through income.

$$\text{Rate of capital turnover} = \frac{\text{Gross income on the farm}}{\text{Total value of farm assets}} \quad 100$$

6.4. Farm Records

Records are statements of fact or data concerning a specific subject which may be specified in physical, monetary, mathematical or statistical terms. Farm records pertain to information recorded on the day-to-day operation of a particular farm.

Farm records can the defined *as* ***systematic documentation of all activities taking place in a farm enterprise over a given period of time. It is an act of writing down every activity engaged in on the farm in every production season and at different stages of the production process up to the final disposal of the goods and services to the ultimate consumer***.

Farm record keeping is more than just keeping track of what crop was planted in what field, it is a concept applicable to the entire farm operation.

A complete farm record will include all daily activities and transactions and with a proper accounting system it should be possible to have a complete estimate of the profit or loss statement at the end of the year. By keeping a complete farm record on a yearly basis it is possible to have a comparison between years and thereby determine the rate of growth or deterioration of the farm.

There are three basic types of farm records:

1. Resources inventories
2. Production accounts of livestock and crop operations.
3. Income and expenses records.

As a farmer, you use resources such as land, labour, machinery, breeding stock management and financial capital. You must assign value to these resources and maintain current inventories so that you know the foundation from which you operate, from this resource base, you produce food and fibre. The production account of livestock and crop shows how this resource base is combined with other inputs to yield physical output. Consequently, production records, such as the relationship of feed to weight gain, cow to number of calves, fertilizer to crop yield, or land to total production, are most critical to your farm's profits. Because farmers interact in a marked economy, the cost of the feed and revenue from the market hog or the cost of the fertilizer and the price of the grain are equally important.

6.4.1. Main Reasons For Keeping Farm Records

1. **The requirement of maintaining with lenders, agencies, etc.:** Lenders, government agencies, insurance companies, and several other bodies demand good records before providing help to farmers. Extension Agents

of Agriculture Agencies sometimes want good records kept by farmers to be able to give them good technical advice on their farms. Financial institution and other lenders would want good records on the farm's income and expenditure before giving out loan facilities to farmers.

2. **Better farm planning and forecasting:** Farming is a business and excellent farm records keeping helps the farmer to plan and do realistic forecasting. A record keeping provides valuable information on the techniques that work and why others did not. In budgeting, the farmer can better predict price changes of inputs and make from expenditure and sales records kept from previous seasons. The farmer can determine how much is necessary for farm expansion from good farm records.
3. **Better prediction of rainfall and weather patterns:** A record on rainfall incidence and amount over a period of time is very relevant. A farmer, from the records, decides when to plant and when to take out other activities on the farm. More so, farmers who depend solely on rainfall, use these records as a guide during the farming season. Besides, temperature changes have significant effects on plants, especially at tender stages. It is, therefore, important for farmers to have a fair thought on temperature changes at certain times in the season. We advise farmers to keep an easy rain gauge and thermometer for the measurements of rainfall amount and temperature respectively.
4. **Track income and expenditure:** A farmer most of the time is not able to give proper accounts on their cash inflows and outflows. This is mainly because they overlook and do not record "petty" expenses and incomes (like feeding and costs of phone calls). Keeping accurate records will at any time of reconciliation, give the farmer the accurate amount of money spent or gained from the farm. Moreover, this helps in accurate planning and budgeting.
5. **Better management of the farm:** A farmer who keeps records on seed germination rates of the seeds bought is in a better position to choose better seeds for seasons. Besides, records keeping help to decide the right type of crop to grow on a particular type of soil, at a particular time of the season for best yield and better pest and disease management. In the case of livestock, the farm keeps farm records of the bloodlines, pest, and diseases, feed types and consumption, etc. These records help to stop inbreeding, control pests and diseases, supply the best feed for optimum performance and a whole lot.

Farmers cannot continue to avoid this activity for the cause of time consumption or any other. So, we advise that spending an average of ten minutes each day on the farm to keep farm records is very much worth it. It will save the farmer time and money.

6.4.2. Importance of Farm Record

1. Farm records help a farmer to keep stock and manage each feature of the farm properly.
2. They are very important for planning and budgeting. They provide a farmer with enough information needed for good planning and budgeting at every point in time.
3. It helps the farmer to get loans and subsidies from the government.
4. They help farmers recognize the progress and contributions of each aspect of the farm to its overall success.
5. They are important for accurate farm management.
6. It helps the farmers to create budgets and plan his business.
7. They can be helpful when a farmer needs to access financial aids from banks or other financial institutions.
8. It enables farmers to create better management and decision.

6.4.3. Various Types of Farm Records Needed to Maintain on Farm

1. **Daily farm records:** These are the records of all important daily activities and events that occur on the farm. These records help the farmer maintain track of past farming activities and plan for future activities.
2. **Records of farm implements and equipment:** This is used to maintain an inventory of all the equipment on the farm and their quantity. It can also contain the date of purchase of the equipment and their description.
3. **The cash book and payment receipt:** This is one of the types of farm records a good farmer must take note, cash book, and payment receipt is a farm record book of all financial transactions, both income and expenditure, of the farm. With it, a farm manager can access loans, detect fraudulent practices and identify the state of the business. It is a farm record accounting book. In addition, the farm manager can create productive decisions based on the figures present in the record.
4. **Record of agricultural inputs:** This record is used to keep track of all agricultural inputs such as fertilizers, seeds, and etc. The record often contains the amount of that was bought, the amount that has been used, and what is left.

5. **Records of livestock and livestock products:** Farmers keep different records of livestock for every type of livestock on the farm. Also, for livestock that have products, for example, eggs from chickens, the farmer keeps a record that accounts for the number of eggs laid every day. If a farmer has cows for milk production, he keeps records of the cows gave how much amount of liters of milk produced per day.
6. **Annual valuation record:** It is imperative that at the end of every production year, a farmer should evaluate his or her venture to know the current state of the business. The farm strength and weakness have to be found; this is possible when the farmer has the annual valuation record. This farm record keeping book shows the value of stocks presents on the farm from the beginning of the year to the end. It will aid decision making to prepare the farm arrangement for the next production year.
7. **Records of animal feeds:** This record is used for maintaining an inventory of the types of animal feed and the quantity purchased, used, and quantity in stock on a daily basis.
8. **Production records:** Production records are used to document everything that is formed on the farm. These production records are prepared every week. And then summed up at the end of the month and at the end of the year. Thus, there is a weekly record, a monthly record, and an annual record of everything formed on the farm.
9. **Records for farm use:** Used for recording the date the land was arranged for farming, the number of plots or hectares used in planting, the plants planted on the farm and where they are planted.
10. **Worker's records:** Worker's record is used to keep the record of staffs, their salaries, and payment. It is also known as labor record.
11. **Sales record:** Sale records are used to keep a record of all sales completed from the farm produce.
12. **Yield and production record:** Read: Plant Disease Management: This is one of the necessary farm records to be kept on a farm; it is the farm record that explains the yield or the output of the farm. For both animal and crop production; the daily or seasonal production rate records in this record. It explains the success rate of the farm during the production year.
13. **Farm input utilization record:** This farm record shows all the input acquired and utilized during the production year, it shows the relevance of a particular input in the production. It shows the level of input application and it is used in combination with output record to determine resource productivity.

14. **Profit and loss record:** This is the principal record; it shows the economic stage of the business. This farm accounting record is the record most farmers tend to keep; however, the aforementioned records are prerequisite to this record. This farm record book is generally prepared at the end of every production period, usually at the end of the year to verify whether the business is making a profit or otherwise. These are the different types of farm records an aspiring successful farmer must prepare and keep; they are computer-aided farm records. The importance of farm records are copious and cannot be overruled; the uses of farm record:
    - Procurement of a bank loan.
    - Aids productive decision making.
    - Help to detect fraudulent practices.
    - It helps to know the economic status of the farm.
    - It shows the strength and weakness of a farm.

6.5. Farm Inventory

The list of all the physical property of a business along with their values at a specific point of time is called **farm inventory**. It is the complete list of farmer's assets. It is the first step in farm accounting as it forms the basis for the preparation of income statement, balance sheet, measurers of income, etc.

Inventory for a business is taken at **two point's of time in a year** i.e., at the **beginning** of the agricultural year and at the **end** of the year. It constitutes cash assets, depreciable assets and non-depreciable assets. The difference in the inventory at the two points of time indicates the changes in the inventory.

The closing inventory for a year is the opening inventory for the next year.

Importance or Purposes

1. The farm inventory is a necessary step in complete farm accounting.
2. It provides information about all the assets with their values.
3. It shows what amount of capital accumulation goes back into the business.
4. It reveals the changes in assets through comparison of two inventories.
5. It enables to determine the depreciation costs.
6. It forms the basis of income statement.

7

Farm Planning

Abstract

This chapter examines how farm planning can help farmers anticipate and optimize their future activities based on economic principles. It presents a comprehensive farm plan that outlines the enterprises, procedures, labour, investments and other information involved in the farm operation. It demonstrates how farm planning can assist farmers in achieving their objectives, choosing crops, allocating resources, estimating input and credit requirements, and projecting future costs and returns.

7.1. What is Farm Planning?

- Farm planning is the foundation of management.
- Farm planning precedes all other managerial functions.
- It is the determination of a course of action to achieve the desired results.
- It is deciding in advance, viz., what to produce?, how to produce?, when to produce?; how to borrow?, how much to borrow?, when to borrow?, where to buy and sell?, when to buy and sell?, how to buy and sell?, etc.
- Farm planning bridges the gap from where we are to where we want to go.
- It is characterized as the process of thinking before doing.
- It is an intellectual process.
- It governs the survival, progress and prosperity of farm organization in a competitive and dynamic environment.
- It is a continuous and unending process.

7.2. Informal and Formal Planning

- Farm planning is not a new technique. It is as old as farming itself. All farmers whether progressive or backward, literate or illiterate, plan their farm business at the beginning of every crop season in their minds. This is known as **informal planning**.
- As the agriculture is becoming a more complex business, the farmers cannot remember all the necessary information required in order to plan

their farm business in a systematic manner. To be more systematic in thinking, the farmers need formal or written farm plans. If the production intentions of the farmers are committed on a paper, then it is called a **formal or written plan**.

7.3. Definition

- **Farm Planning** can be defined as the deliberate and conscious effort on the part of the farmer to think about the farm programmes **in advance** and adjust them according to new knowledge on the technological developments, changes in physical and economic situations, price structures, etc.
- **Farm planning** is a process of making decisions regarding the organization and operation of farm business, so that it results in continuous maximum net returns of farm business.

7.4. Why Farm Planning is Necessary?

It has been observed on majority of farms that there is **under or over utilization** of existing farm resources. Due to this, farmers fail to get maximum net return. Thus, it is immediate necessity of a farm to reorganize its structure and for the proper allocation of its resources. This requires proper farm planning.

7.5. Importance of Farm Planning

1. To choose different farm activities which are suited to the given farm conditions.
2. To look into the future and decide on suitable course of action.
3. To select appropriate enterprise combinations that results in the better resources.
4. To help the farmers in timing various jobs and operations for smooth conduct of operations without competition.
5. To avoid wastages that occur in the resource use.
6. To provide guidance and flexibility to the farmers for ensuring better use and growth of the farm.
7. To provide appropriate allocation of resources for producing the requisite products for marketing and household consumption.

7.6. Objectives of Farm Planning

1. Efficient use of scare resources.
2. Selection and combination of different farm enterprises or activities.
3. Utilization of advance technology.

4. Easy and efficient management of farm business.
5. Continuous and maximum net returns of farm business.

7.7. The Information Required for Farm Planning

1. Financial returns of farm
2. Requirements of variable inputs
3. Requirements of fixed resources

7.8. Types of Farm Planning

- **Simple farm planning:** It implies planning for minor changes or for a particular enterprise. It is adopted either by a part of land for one enterprise or to substitute one resource by another.
- **Complete farm planning**: It envisages farm planning for the whole farm, i.e. for all enterprises on the farm, for a change in the farm structure and organization. Complete farm planning aims for a change in the entire farm structure and organization. It is adopted for the farm as a whole.

7.9. What is a Farm Plan?

A farm plan is a programme of the total farm activity drawn up in advance.

17.10. Characteristics of Good Farm Plan

1. It should provide **for efficient use of farm resources** such as land, labour and capital.
2. It should provide **balanced combination** of enterprises.
3. It should provide **appropriate combination** of different crops.
4. It should help to maintain and improve **soil fertility**.
5. It should help to raise and stabilize farm **earnings**.
6. It should avoid excessive **risks**.
7. It should provide **flexibility**.
8. It should utilize the farmer's **knowledge**, training **experience** and take into account of the farmer's **likes and dislikes**.
9. It should provide for all of **up-to-date** modern agricultural methods and practices.
10. It should give considerations to efficient **marketing facilities**.
11. It should provide programme of obtaining, using and repaying the **credit**.

7.11. Basic Steps of Farm Planning

1. Statement of objective.
2. Survey of actual conditions of the farm and availability of resources.
3. Analysis / diagnosis of the existing plan.
4. Identification of enterprises to be included.
5. Discussion with the farmers and other specialists to examine the possibilities for improvement through alternative plan operations.
6. Preparation of alternative plans on the basis of existing co-efficient.
7. Identification of risk.
8. Selection of final plan for implementation.

7.12. Farm Planning Techniques

a) Production and organization planning tools

Production function models,	Linear programming,
Operational research techniques,	Dynamic programming,
Integer programming,	Recursive programming
Modified programming models	Game theory models, etc.

b) Administrative planning tools and aids

Farm technical coefficients,	Accounting & data processing,
Management consulting agency,	Contractual farming or operations, etc.

c) Marketing planning tools and aids

Demand supply projection analysis	Future trading analysis
Marketing management research,	Price policies, etc.

7.13. Limitations of Farm Planning

1. It is time consuming and expensive device.
2. It encourages a false sense of security against risks.
3. Unavailability of adequate information.
4. Inaccurate or false information.
5. The information particularly in respect of climate, water supply, markets, etc., is not found in the required form.
6. The sources of data for diagnosis and planning are lacking.

As a result of above limitations, farm planning is not effectively formulated and implemented. Therefore, farm standards derived from research stations and efficient farms in the locality should form the basis for scientific planning. Data from research stations should be continuously used for this purpose.

8

Farm Budgeting

Abstract

This chapter explores farm budgeting, a method of evaluating plans for the allocation of agricultural resources at the decision-maker's level. It involves estimating the income, costs, and profit of different farm activities.

8.1. What is Farm Budgeting?

- The expression of a farm plan in **monetary terms** by estimation of **returns, expenses** and **net income** is called **farm budgeting**.
- Farm planning and farm budgeting **go side by side**.
- It is essentially a presentation of **costs and returns**, accompanied by a statement showing the **physical quantities of inputs and output**.

8.2. Objectives of Farm Budgeting

1. **To estimate the profitability** of a particular pattern of organization
2. **To determine the change in profits** which is likely to follow a particular change.
3. **To compare** different cropping patterns or changes in organization **on a profit basis**.

8.3. Types of Farm Budgeting

1. Enterprise budget
2. Partial budgeting
3. Complete budgeting

8.4. Farm Enterprise Budget

- It is the estimation of expected **income, costs and profit** for an **enterprise.**
- It is prepared **for a unit** i.e., acre, hectare, for a crop, one head of livestock, etc.
- Farm budgets can be developed **for each potential enterprise**.
- This facilitates easy **comparison** between the enterprises

How to prepare an enterprise budget?

1. It consists of three elements viz., **income, costs and profitability**.
2. **Income** is computed by estimating the expected output and expected price.
3. The estimated **output** is the average yield under normal weather conditions.
4. **Output price** should be the average price expected in future.
5. **Variable costs** should be estimated on the quantity of inputs required and their prices.
6. **Fixed costs** to be included such as land revenue, depreciation, interest on fixed capital and rental value of d land, etc.
7. Finally, **net profit** is to be estimated on bases of estimated income and expenditure.

Table 18.4: Enterprise Budget for Pea

S.No.	Particulars	Per bigha			Per ha	
		Quantity	Rate (₹)	Value (₹)	Quantity	Value (₹)
A	**Variable cost**					
1	Seed (kg)	22	33	726	275	9075
2	Seed Treatment			18		225
3	FYM (q)	7.50	162	1215	93.75	15187.5
4	Fertilizer					
I)	IFFCO mixtur (kg)	13	10.70	139.10	162.50	1728.75
II)	Urea (kg)	6	6	36	75	450
5	Plant protection			418		5225
6	Bullocklabou(days)	2	500	1000	25	12500
7	Human Labou (mandays)		150	450		
I)	Field prepration	3	150	450	37.50	5625
II)	Seed preparation & Sowing	3	150	450	37.50	5625
III)	Manuring	3	150	1650	37.50	5625
IV)	Interculture	11	150	600	137.50	20625
V)	Irrigation	4	150	450	50	7500
VI)	Spraying	3	150	1050	37.50	5625
VII)	Harvesting/Picking, packaging & transportation	7	150		87.50	13125
VIII)	Total human labour of which	34		5100	425	63750

I)	Family labour	22	150	3300	275	41250
II)	Hired labour	12	150	1800	150	22500
8	Sub total			8652.10		108151.3
9	Interest on working capital			115.36		1442.02
10	Total variable cost (A=8+9)			8767.46		109593.3
B	**Fixed cost**					
I)	Rental value of land			5000		62500
II)	Interest on fixed capital			648		8100
III)	Depreciation			810		10125
	Total fixed cost			6458		80725
C	Total cost (A+B)			15225.46		190318.3
D	Production and Returns					
1	Production (q)			10.75		134.375
2	Average price (₹/kg)			22		22
3	By product (q)			1.02		12.75
4	Average price (₹/kg)			15		15
5	Gross Return			25180		314750
6	Net Return (₹)			9954.54		124431.70

8.5. Partial Budgeting

- Partial budgeting is **a statement of anticipated changes** in costs, returns and profitability for a minor modification.
- When a farmer contemplates few **modifications or minor changes** in the existing plan of the farm business, partial budgeting technique is employed.
- It is similar to that of **marginal analysis**, wherein the changes in costs and returns resulting from proposed modifications are alone considered.
- It is prepared in the **form of an account** which consists of **four important elements** viz., added costs, added returns, reduced returns and reduced costs.
- It is generally used to evaluate the profitability of **input substitution, enterprise substitution and scale of operation.**

Elements of Partial Budgeting

Debit	Credit
Added costs : Additional costs are incurred, if the proposed modification is the introduction of a new enterprise or increase in the size of the existing enterprise.	**Added returns** : Additional returns could be received when the proposed modification is the addition of a new enterprise, or increase in the size of the existing enterprise or adoption of technology that results in higher productivity.
Reduced returns : Decrease in the returns is observed when the proposed modification involves the elimination of an existing enterprise or reduction in the size of the existing enterprise.	**Reduced costs** : Decrease in the costs is found when the proposed modification involves the elimination of existing enterprise or reduction in the size of the enterprise or adoption of a technology that uses fewer amounts of resources.
Total Debit	**Total Credit**
Net Income from the change (Total Credit – Total Debit) :	

8.6. Complete Budgeting

- This method is used to make a budget **for the whole farm**.
- It is used in situations involving **extensive remodeling** of the farm organizations.
- It includes all the individual cost and return items for the farm, so that the overall net return can be estimated **from the whole farm**.
- It considers **all aspects** of farm organization **simultaneously**.

Steps in farm planning and budgeting: The sound farm plan should be generally feasible, acceptable, and adaptable. To make the farm plan successful, the following steps should be adopted with relevance to given farm and its resources.

1. Statement of objective.
2. Diagnosis of the existing organization
3. Assessment of resource endowments on the farm.
4. Identification of enterprises to be included.
5. Preparation of enterprise budgets.
6. Identification of risks, and
7. Preparation of a plan.

1. **Statement of objective**: The objective of the farmer may be profit maximization or cost minimization. In selecting enterprises and their

combinations, the farmer aims at maximization of profits. On the other hand, while choosing resources and their combinations, he aims at cost minimization.

2. **Diagnosis of the existing organization:** Diagnosis and prescription are the two important components of planning. The planner has to examine the existing organization of farm business carefully and identify the weaknesses or defects or loopholes in the current plan. Once mistakes are identified, corrective steps can be taken in future. Farm plans primarily prescribe remedies for the defects of the existing plan.

3. **Assessment of resource endowment on the farm:**

a) **Land:** Here there is a need to spell out the land holding area, type of land i.e. wet land or dry land, crops grown, type of soils available, topography, texture, fertility status, drainage, soil and water development, soil and water conservation methods, etc.

b) **Labour:** The extent of family labour available with the farmer viz., women, men and children along with their age, household work and farm work done by them should be indicated. Permanent labourers if any engaged by the farmer, type of work done and amount of remuneration paid should be indicated. Labour supply, in the village and demand for labour for different crops in different seasons should be assessed. The supply position with reference to livestock should be assessed correctly.

c) **Capital:** Working capital required for raising crops should be indicated. Owned funds available and the amount of funds borrowed, from different sources, interest paid, etc., need to be clearly specified. Specification of repayment dates, terms and conditions, etc., is also required. Fixed capital relates to information on farm buildings, farm equipment, farm machinery, etc.

d) **Organization:** The farmer's knowledge in farming, his expertise, his experience in farming and confidence in adapting new potential technology should be assessed. Based on this information relevant farm plan should be devised. If the farmer is risk-averse, farm plans, which provide stable income under risk, should be generated.

e) **Irrigation sources:** Availability of different sources of irrigation, area covered under different sources, period of availability of irrigation, quantity of irrigation water available, crop demands for irrigation water, accessibility of land to the irrigation sources such as canal and tank, etc., should also be indicated. In addition to this cost of irrigation needs to be mentioned.

4. **Identification of enterprises to be included:** List of enterprises not only grown by the farmer but also enterprises grown in that area and also crop rotations are identified. Estimate the input-output coefficients in terms of acre or hectare or head of livestock for all the enterprises, which we propose to include. Information on input and output prices should be collected so as to work out the costs and returns.
5. **Preparation of enterprise budgets:** Estimate the income, cost and profitability of each enterprise to be included in the plan. The preparation of enterprise budgets facilitates comparison of profitability of different enterprises.
6. **Identification of risks:** List out all types of risks viz., production risk, weather risk, technological risk, institutional risk, marketing risk, etc., faced by the farmers. Particularly the incidence of pests, rodents and diseases, frequency of drought occurrence over time, cyclones, floods and their havoc caused to farm plans. Marketing risks comprising of risk emanating from price fluctuations and failure of markets to arrest the malpractices of middlemen should be indicated.
7. **Preparation of a plan:** The first step is identifying the scarcest resources and selecting that enterprise which yields maximum returns per unit of scarcest resource. This process is repeated till all the scarce resources are put to the best use which results in optimum combination of the enterprises.

Difference between partial budgeting and complete budgeting

S.No.	Partial Budgeting	Complete Budgeting
1.	It consider few alternatives	It consider all alternatives
2.	Do not affect the organization vitally	Affect the organization
3.	It is adopted when drastic changes in the existing organization are contemplated	Adopted when minor changes are introduced on the farm.
4.	Easy and simple it require less time and efforts	It require more time efforts and more data in accurate form

8.7. Example 1 : Partial Budgeting

For a farmer growing groundnut, the expenditure on seeds and groundnut yield in existing (local variety) and alternate situation (improved variety) are as under:

Existing situation	Alternate situation
Local variety: 50 kg @ Rs. 30/kg = Rs. 1500	Improved variety: 50 kg @ 50/kg = Rs. 2500
Yield= 600 kg @ Rs. 30/ kg. = Rs.18000	Yield=700 kg @ Rs. 30/kg.= Rs.21000

The partial budget for proposed modification

Items	Debit	Items	Credit
Added costs:	Rs. 1000	**Added returns:**	Rs. 3000
Reduced returns:	Nil	**Reduced cost:**	Nil
Total Debit (added costs + reduced returns)	**Rs. 1000**	**Total Credit** (added returns + reduced cost)	Rs. **3000**
Net income from the change = Total Credit – Total Debit = Rs. 3000 – Rs. 1000 = Rs. 2000			

8.8. Example 2: Enterprise Budget

Enterprise Budget for cotton per Hectare

S.No.	Particulars	Physical Unit	Rate	Amount Rs.
1	**Income**			
	Yield	25 q	5500/q	**1,37,500**
2	**Expenditure**			
	(a) Fixed Cost			
	Rental value of land	1 ha	30000	30,000
	Interest of fixed capital	-	-	1,000
	Depreciation	-	-	3,000
	Land revenue	-	-	1,000
	Sub-Total			**35,000**
	(b) Variable cost			
	Labour	200	100	20,000
	Seeds	1.2 kg	2500/kg	3,000
	Bullock labour	20	300	6,000
	Tractor charges	-	-	5,000
	Fertilizers	-	-	10,000
	Irrigation	-	-	6,000
	Plant Protection chemicals	-	-	5,000
	Miscellaneous	-	-	5,000
	Sub-total			**60,000**
	Total expenditure			**95,000**
3	**Net Profit**			**42,500**

9

Linear Programming

Abstract

This chapter covers how linear programming aids in reducing costs and maximizing gains. A linear programming problem (LPP) typically contains linear restrictions and an objective function seeking optimization. Initially, two-variable linear optimization models are examined using the graphical technique to comprehend the issue and investigate solutions. The graphical method provides visual insights into the problem's nature and solutions. Linear programming identifies the mix of choices that optimizes the objective function subjected to constraints. Graphical interpretation aids in recognizing feasible and optimal solutions visually.

9.1. Origin of Linear Programming

All business man or industrialist has to find out of all permissible allocation of variable resources on which optimize the total cost or profit (i.e. minimization the cost and maximization the profit). The use of linear programming has been found to be most useful in such decision making problems.

The Russian mathematician **L.V. Kantorovich** first of all applied mathematical methods to solve such problems. He pointed in **1939** that many classes of problems which arise in production can be defined as mathematically and therefore can be solved numerically.

The decision-making technique was later on developed by **George B. Dantzing** and he formulated the general linear programming problems and developed the simplex method in **1947** and the ***theory of duality was developed by John von Neumann*** the same year (i.e. 1947) for solving such problems.

It was first applied in 1930 by economist, mainly in solving resource allocation problem During World War II, the US Air force sought more effective procedure for allocation of resources George B. Dantzig, a member of the US Air Force formulate general linear problem for solving the resources allocation problem.

9.2. Definitions of Linear Programming

1. It is an optimization method applicable for the solution of optimization problem where objective function and the constraints are linear.

2. Linear programming is a budgeting technique that is more refined and systematic than the conventional budgeting method. It is a technique that uses computational procedure.
3. LP is a resource allocation model that seeks the best allocation of limited resources to a number of competing activities.
4. Linear programming (LP), also called linear optimization, is a method to achieve the best outcome (such as maximum profit or lowest cost) in a mathematical model whose requirements are represented by linear relationships.
5. In simple terms, it is a method to find out how to do something in a best possible way. In other words, given limited resources you need to do optimum utilization of resources to achieve the best possible outcome. i.e. maximizing profit, minimizing cost
6. **Dorfman *et al.*,** define linear programming as "the analysis of the problems in which linear function of a number of variables is to be maximized (or minimized) when those variables are subject to a number of restraints in the form of linear inequalities."
7. "A Linear Programming Problem is one that is concerned with finding the optimal value (maximum or minimum value) of a linear function (called objective function) of several variables (say x and y), subject to the conditions that the variables are non-negative and satisfy a set of linear inequalities (called linear constraints). The term linear implies that all the mathematical relations used in the problem are linear relations while the term programming refers to the method of determining a particular programme or plan of action."
   - It solve the problems of decision making.
   - LP is mathematical modelling technique in which objective functions maximized and minimized.
   - It is helpful to guiding quantitative decision in particular business planning.
   - It use to achieve optimality point (solutions) and determine the optimal solutions.

9.3. Assumptions of Linear Programming

1. ***Proportionality:*** Linear programming assumes that any modification in the constraint inequalities will result in a proportional change in the objective function. This means that if it takes 10 hours to produce 1 unit of a product, then it would take 50 hours to produce 5 such products.

2. ***Additivity:*** Additivity means that each function in a linear programming model is the sum of the individual contributions of the respective activities. For example, the total profit is determined by the sum of profit contributed by each activity separately.
3. ***Divisibility:*** Linear programming makes the divisibility assumption that the solution has to be in whole numbers i.e. integers. This assumption means that decision variable may take any value, including non-integer values, as long as functional and non-negativity constraints are satisfied.
4. ***Certainty:*** Certainty in linear programming refers to the assumption that the parameters of the objective function coefficients and the coefficients of constraints are known with certainty. For example, profit per unit of product, resource availability per unit, etc. are known with certainty.
5. ***Non-negativity:*** Linear programming assumes that all answers or variables are non-negative. This assumption is true in the sense that negative values of physical quantities are not possible. It is not possible for the output in the production problem (such as bicycles, cars, computers, etc.) to be negative.
6. ***Optimality:*** In linear programming problems of maximum profit solution or minimum cots solution always occurs at a corner point of the set of the feasible solution.
7. ***Finiteness of activity and resource restrictions:*** Linear programming assumes about the presence of a finite number of activities. An optimal solution is not possible in a situation where there is an infinite number of alternative activities and resource constraints.
8. ***Continuity:*** Linear programming assumes the continuity of decision variables. This means that a combination of outputs with fractional values plus integer values can be used.

9.4. Properties of Linear Programming Model

Linear programming model has the following conditions:

- The decision variables must have a linear relationship.
- It must have an objective function.
- Resource constraints are required.
- It must have a non-negative constraint.

A linear programming model involves an objective function, well-defined decision variables, and a set of non-negative structural constraints.

Let us try to understand these terms in the following section:

- Objective Function
- Decision Variables
- Constraint Equation
- Structural Constraints
- Non-negativity Constraint

9.5. Concept of Slack, Surplus and Artificial Variable

When decision variables are more than 2, we always use simplex method.

9.5.1. Slack Variable

- Variable added to a ≤ constraint to convert it to an equation (=).
- Slack variable represents unused resources.
- A slack variable contributes nothing to the objective function value. j = 1.
- For determination of feasible solution, each inequality in the constraint equation be written as equality for the purpose of drawing straight line.
- Eg. $2X_1 + 3X_2 + S_3 = 60$ Inequality is converted in to equality. Thus, slack variable is used to convert less than

The Simplex Method: Maximization (Adding Slack Variable)

Suppose we want to find the maximum value of where and subject to the following constraints.

$-X_1 + X_2 \leq 11$

$X_1 + X_2 \leq 27$

$2X_1 + 5X_2 \leq 90$

Since the left-hand side of each inequality is less than or equal to the right-hand side, there must exist nonnegative numbers S_1, S_2 & S_3 that can be added to the left side of each equation to produce the following system of linear equations.

$- X_1 + X_2 + S_1 = 11$

$X_1 + X_2 + S_2 = 27$

$2X_1 + 5X_2 + S_3 = 90$

The numbers S1, S_2 & S_3 are called slack variables because they take up the "slack" in each inequality.

9.5.2. Surplus Variable

- Variable subtracted form a $\geq$ constraint requirement level.
- A surplus variable represents the amount by which solution values exceed a resource. These variables are also called 'Negative Slack Variables'
- Surplus variables like slack variables carry a zero coefficient in the objective function. it is added to greater than or equal to (>) type constraints in order to get an equality constraint.
- A variable which is subtracted from the left-hand side of a greater than or equal to constraint in to equality is called surplus variable.

$a_{11}X_1 + a_{12}X_2 + \ldots\ldots\ldots\ldots a_{1n}X_n - X_{n+1} = b_1$

Example

$2X_1 + 3X_2 \geq 60$

$2X_1 + 3X_2 \geq 60 + X_3$

$2X_1 + 3X_2 - X_3 = 60$

X_3 is Surplus variable

9.5.3. Artificial Variable

- Artificial variables are added to those constraints with equality (=) and greater than or equal to (>) sign.
- A fictitious variable included in the case of inequality of $\geq$ type and equality (=), to satisfy the non-negativity condition of a basic feasible solution.
- Introduce an Artificial Variable in each = constraint.
- For each artificial variable A, add – A to the objective function.
- Use the same constant M for all artificial variables.

Objective function

Max $Z = 3X_1 - 2X_2 + X_3$

Subject to

$X_1 - 2X_2 + X_3 \geq 5$

$- X_1 - 3X_2 + 4 X_3 \leq - 10$

$2X_1 + 4X_2 + 5 X_3 \leq 20$

$3X_1 - X_2 - X_3 = -15$

Non negative constraint

$X_1, X_2, X_3 \geq 0$

Introduce a surplus variable and an artificial variable for the first constraint:

$X_1 - 2X_2 + X_3 \geq 5$

$X_1 - 2X_2 + X_3 - S_1 + A_1 = 5$

Notice that the second constraint has a negative number on the right-hand side. To make that number positive multiply both sides by -1 and reverse the direction of the inequality:

$X_1 + 3X_2 - 4X_3 \geq 10$

The fourth constraint has a negative number on the right-hand side so multiply both sides of this equation by -1 to change the sign of - 5 to + 15.

$-3X_1 + X_2 + X_3 = 15$

Do the same procedure for the Second constraint: ≥

$X_1 + 3X_2 - 4X_3 - S_2 + A_2 = 10$

Introduce surplus variable for less than or equal to third constraint:

$2X_1 + 4X_2 + 5X_3 + S_3 = 20$

Introduce the third artificial variable for the equation constraint:

$-3X_1 + X_2 + X_3 + A_3 = 15$

For each of the three artificial variables, we will add - A to the Objective function.

$Z = 3X_1 - 2X_2 + X_3 - A_1 - A_2 - A_3$

Final result, Modified problem is,

$\text{Max } Z = 3X_1 - 2X_2 + X_3 + 0S_1 + 0S_2 + 0S_3 - MA_1 - MA_2 - MA_3$

Subject to the constraints

$X_1 - 2X_2 + X_3 - S_1 + A_1 = 5$

$X_1 + 3X_2 - 4X_3 - S_2 + A_2 = 10$

$2X_1 + 4X_2 + 5X_3 + S_3 = 20$

$-3X_1 + X_2 + X_3 + A_3 = 15$

$X_1, X_2, X_3 \geq 0$

9.6. Linear Programming Formula

A linear programming problem will consist of decision variables, an objective function, constraints, and non-negative restrictions. The decision variables, x, and y, decide the output of the LP problem and represent the final solution. The objective function, Z, is the linear function that needs to be optimized (maximized or minimized) to get the solution. The constraints are the restrictions that are imposed on the decision variables to limit their value. The decision variables must always have a non-negative value which is given by the non-negative restrictions. The general formula of a linear programming problem is given below:

- **Objective Function:** $Z = ax + by$
- **Constraints:** $cx + dy \leq e$, $fx + gy \leq h$. The inequalities can also be "≥"
- **Non-negative restrictions:** $x \geq 0$, $y \geq 0$

How to Solve Linear Programming Problems?

The most important part of solving linear programming problem is to first formulate the problem using the given data. The steps to solve linear programming problems are given below:

- **Step 1:** Identify the decision variables.
- **Step 2:** Formulate the objective function. Check whether the function needs to be minimized or maximized.
- **Step 3:** Write down the constraints.
- **Step 4:** Ensure that the decision variables are greater than or equal to 0. (Non-negative restraint)
- **Step 5:** Solve the linear programming problem using either the simplex or graphical method.

Let us study about these methods in detail in the following sections.

9.7. Linear Programming Methods

There are two main methods available for solving linear programming problem. These are the simplex method and the graphical method. Given below are the steps to solve a linear programming problem using both methods.

9.7.1. Linear Programming by Simplex Method

The simplex method in lpp can be applied to problems with two or more decision variables. Suppose the objective function $Z = 40x_1 + 30x_2$ needs to be maximized and the constraints are given as follows:

$x_1 + x_2 \leq 12$

$2x_1 + x_2 \leq 16$

$x_1 \geq 0, x_2 \geq 0$

Step 1: Add another variable, known as the slack variable, to convert the inequalities into equations. Also, rewrite the objective function as an equation.

$- 40x_1 - 30x_2 + Z = 0$

$x_1 + x_2 + y_1 = 12$

$2x_1 + x_2 + y_2 = 16$

y_1 and y_2 are the slack variables.

Step 2: Construct the initial simplex matrix as follows:

$$\begin{bmatrix} x_1 & x_2 & y_1 & y_2 & Z & \\ 1 & 1 & 1 & 0 & 0 & 12 \\ 2 & 1 & 0 & 1 & 0 & 16 \\ -40 & -30 & 0 & 0 & 1 & 0 \end{bmatrix}$$

Step 3: Identify the column with the highest negative entry. This is called the pivot column. As -40 is the highest negative entry, thus, column 1 will be the pivot column.

Step 4: Divide the entries in the rightmost column by the entries in the pivot column. We exclude the entries in the bottom-most row.

12 / 1 = 12

16 / 2 = 8

The row containing the smallest quotient is identified to get the pivot row. As 8 is the smaller quotient as compared to 12 thus, row 2 becomes the pivot row. The intersection of the pivot row and the pivot column gives the pivot element.

Thus, pivot element = 2.

Step 5: With the help of the pivot element perform pivoting, using matrix properties, to make all other entries in the pivot column 0.

Using the elementary operations divide row 2 by 2 (R_2 / 2)

$$\begin{bmatrix} x_1 & x_2 & y_1 & y_2 & Z & \\ 0 & 1/2 & 1 & -1/2 & 0 & 4 \\ 1 & 1/2 & 0 & 1/2 & 0 & 8 \\ -40 & -30 & 0 & 0 & 1 & 0 \end{bmatrix}$$

Now apply $R_1 = R_1 - R_2$

$$\begin{bmatrix} x_1 & x_2 & y_1 & y_2 & Z & \\ 1 & 1 & 1 & 0 & 0 & 12 \\ 1 & 1/2 & 0 & 1/2 & 0 & 8 \\ -40 & -30 & 0 & 0 & 1 & 0 \end{bmatrix}$$

Finally, $R_3 = R_3 + 40\,R_2$ to get the required matrix.

$$\begin{bmatrix} x_1 & x_2 & y_1 & y_2 & Z & \\ 0 & 1/2 & 1 & -1/2 & 0 & 4 \\ 1 & 1/2 & 0 & 1/2 & 0 & 8 \\ 0 & -10 & 0 & 20 & 1 & 320 \end{bmatrix}$$

Step 6: Check if the bottom-most row has negative entries. If no, then the optimal solution has been determined. If yes, then go back to step 3 and repeat the process. -10 is a negative entry in the matrix thus, the process needs to be repeated. We get the following matrix.

$$\begin{bmatrix} x_1 & x_2 & y_1 & y_2 & Z & \\ 0 & 1 & 2 & -1 & 0 & 8 \\ 1 & 0 & -1 & 1 & 0 & 4 \\ 0 & 0 & 20 & 10 & 1 & 400 \end{bmatrix}$$

Writing the bottom row in the form of an equation we get $Z = 400 - 20\,y_1 - 10\,y_2$. Thus, 400 is the highest value that Z can achieve when both y_1 and y_2 are 0. Also, when $x_1 = 4$ and $x_2 = 8$ then value of $Z = 400$ Thus, $x_1 = 4$ and $x_2 = 8$ are the optimal points and the solution to our linear programming problem.

9.7.2 Linear Programming by Graphical Method

If there are two decision variables in a linear programming problem then the graphical method can be used to solve such a problem easily.

Suppose we have to maximize $Z = 2x + 5y$.

The constraints are $x + 4y \leq 24$, $3x + y \leq 21$ and $x + y \leq 9$

where, $x \geq 0$ and $y \geq 0$.

To solve this problem using the graphical method the steps are as follows.

Step 1: Write all inequality constraints in the form of equations.

$x + 4y = 24$

$3x + y = 21$

$x + y = 9$

Step 2: Plot these lines on a graph by identifying test points.

$x + 4y = 24$ is a line passing through (0, 6) and (24, 0). [By substituting $x = 0$ the point (0, 6) is obtained. Similarly, when $y = 0$ the point (24, 0) is determined.]

$3x + y = 21$ passes through (0, 21) and (7, 0).

$x + y = 9$ passes through (9, 0) and (0, 9).

Step 3: Identify the feasible region. The feasible region can be defined as the area that is bounded by a set of coordinates that can satisfy some particular system of inequalities.

Any point that lies on or below the line $x + 4y = 24$ will satisfy the constraint $x + 4y \leq 24$.

Similarly, a point that lies on or below $3x + y = 21$ satisfies $3x + y \leq 21$.

Also, a point lying on or below the line $x + y = 9$ satisfies $x + y \leq 9$.

The feasible region is represented by OABCD as it satisfies all the above-mentioned three restrictions.

Step 4: Determine the coordinates of the corner points. The corner points are the vertices of the feasible region.

O = (0, 0)

A = (7, 0)

B = (6, 3). B is the intersection of the two lines $3x + y = 21$ and $x + y = 9$. Thus, by substituting $y = 9 - x$ in $3x + y = 21$ we can determine the point of intersection.

C = (4, 5) formed by the intersection of $x + 4y = 24$ and $x + y = 9$

D = (0, 6)

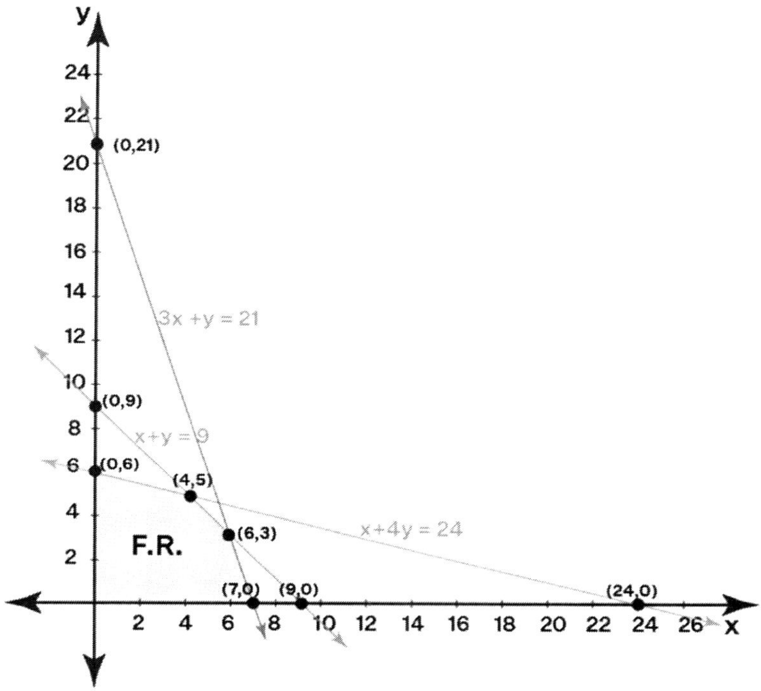

Step 5: Substitute each corner point in the objective function. The point that gives the greatest (maximizing) or smallest (minimizing) value of the objective function will be the optimal point.

Corner Points	**Z = 2x + 5y**
O = (0, 0)	0
A = (7, 0)	14
B = (6, 3)	27
C = (4, 5)	33
D = (0, 6)	30

33 is the maximum value of Z and it occurs at C. Thus, the solution is x = 4 and y = 5.

9.8. Applications of Linear Programming

Linear programming is used in several real-world applications. It is used as the basis for creating mathematical models to denote real-world relationships. Some applications of LP are listed below:

- Manufacturing companies make widespread use of linear programming to plan and schedule production.
- Delivery services use linear programming to decide the shortest route in order to minimize time and fuel consumption.
- Financial institutions use linear programming to determine the portfolio of financial products that can be offered to clients.

9.9. Advantages of Linear Programming

There are several advantages of linear programming as mentioned below:

- ***Scientific approach to problem-solving***: LP employs a scientific approach to problem-solving.
- ***Evaluation of all possible alternatives***: Today's environment presents highly complex decision-making problems to organizations which are difficult to solve by the traditional approach.
- ***Optimal utilization:*** LP enables optimal utilization of various prevailing factors of production such as labour, raw materials, equipment, cost, etc.
- ***Helps in re-evaluation:*** LP helps to re-assess a basic plan in case of changing conditions.
- ***Improves the quality of the decision:*** LP helps to improve quality of decisions by incorporating the limitations of the system.
- ***Addresses bottlenecks:*** LP highlights and addresses the problem of bottlenecks in the production process through optimization.
- ***Applicable to diverse problems:*** LP is quite an accommodating mathematical technique and can be adapted to analyze diverse multi-dimensional decision-making problems quite effectively.
- ***Formation of information base:*** LP models can help managers obtain a highly useful information database by the analysis of the many possible alternatives taking into account the existing constraints.

9.10. Disadvantages of Linear Programming

- Assumption of linear relationship
- Constant value of objective and constraint equations
- No scope for fractional value solutions
- Degree of complexity
- Unsuitability for multiple goals
- Lack of flexibility

10

Risk and Uncertainity

Abstract

This chapter discusses the concepts of risk and uncertainty in decision-making. Risk refers to the condition where the decision-maker knows the possible outcomes and their probabilities. Uncertainty refers to the condition where such information is unknown to the decision-maker.

10.1. Introduction

Agricultural production is biological and seasonal in nature and therefore, it confronted with risk and uncertainty conditions. It is a highly difficult to make right decisions when the production environment is risky and uncertain. Due to this the farmers have to consider various management strategies relevant to the risk and uncertainty conditions. If everything in farming and livestock rearing goes with certainty then every farmer becomes a better manager, but this is not the case with real farming situation. Only a few farmers can become efficient, particularly those who could understand risk and uncertainty situations in farming and follow the relevant risk management strategies.

10.2. Risk and Uncertainty: A Contrast

Risk	Uncertainty
It is defined as a, situation when all possible outcomes are known for a given management decision and probability associated with each possible outcome is also known.	Uncertainty situation prevails when all the possible outcomes of events are unknown, then neitherthe probability nor the outcomes are known.
It is measurable through probability concepts.	It is not measurable.

10.3. Sources of Risk

1. Production risk or technical risk
2. Price risk or marketing risk
3. Financial

10.3.1. Production Risk or Technical Risk

- In industrial businesses we have a technical input-output relationship with known quantity of output for a given quantity of input i.e., the production practices are standardized in industrial production. **Such type of relationship does not exist in agricultural production.**
- Output of crops and livestock is subject to change due to weather, disease, insects, weeds and inadequate technology. These factors cannot be predicted accurately and hence results in the variability of the output.
- 1n fact, the yield variations are due to many factors. Some are under control while some others are not under the control of management.
- Thus, the production-risk is due to largely many factors rather than a single factor.
- **Weather risk and technical risk** are the most important components of production-risk.

10.3.2. Price Risk or Marketing Risk

- Prices are determined due to interaction of demand and supply in the market.
- Production of crops and livestock is influenced by prices in the market which are beyond the control of farmers.
- Prices of commodities vary from year to year; season to season and exhibit seasonal variation.
- Trade policy *i.e.,* export import (Exim) policy would have a great impact on price stabilization of commodity.

10.3.3. Financial Risk

- This type of risk increases with increased amount of borrowed money in the farm business.
- Financing the production processes and storing the commodities under scientific management require large amounts of borrowed capital.
- Hence a careful analysis is to be-made considering all types of risks.

10.4. Measuring the Expectation and Variation

In the risky farm environments, the farmers make decisions using some kind of the expectations regarding prices, costs and yields. Let us look into as to how the expectations and variations are worked out and used in the decision-making process.

10.4.1. Working out expected value through weighted averages

Two types of averages are used to work out an expected value. We take the actual past prices (time series data) and yield to find out the **simple average** over the specific time period. Second type of average is called **weighted average**. This method considers larger weights for the more recent data and smaller weights for the remote data.

Working out weighted averages for the prices

Year	Av. Annual price of commodity (Rs./q)	Weight	Price x Weight
2005	1000	1	1000
2006	1200	2	2400
2007	1400	3	4200
2008	1100	4	4400
2009	1000	5	5000
2010	1300	6	7800
Total	7000	21	24800
Simple average	1166.67		
Weighted average	1180.95		

For the given data there is **no price risk** for the commodity, as **expected value is more than simple average.** On the contrary, if there is price risk its expected value would be less than the simple average.

10.4.2. Working out expected value through the concept of probabilities

Another way of finding expectations is to choose the probability value, which is the most likely to occur for the given price or yield or cost. In this procedure, we have to elicit probabilities associated with each possible outcome. The outcome with the highest probability would be selected as the most likely one to occur. This is shown as under:

Working out Expectations through Probabilities

Crop yield (q/ha)	Probability	Probability x Yield
11	0.1	1.1
14	0.4	5.6
10	0.1	1.0
12	0.3	3.6
18 .	0.1	1.8
Mean : 13	**Total : 1.0**	**13.1**

It is clear that the most likely output based on the highest probability is 14 q/ha. Mathematical expectation of 13.1 q/ha is less than the most likely yield of 14 q/ha, but it is more than the average yield *i.e.,* 13 q/ha.

10.4.3. Variability

Two factors *viz.,* the variance and the expected value, in general should form the basis for selecting the best enterprise among alternative risky enterprises. Range is a measure of variability and it refers to the difference between the lowest and the highest possible outcome. Alternative enterprises should be preferred which have got smallest range value, provided their expected values are the same. Similarly alternative enterprises with lowest variance should be selected.

10.5. Measures to Manage Farm Risk

10.5.1 Diversification

- Selection of suitable crop and livestock enterprises is the first step in diversification.
- The farmer produces several products rather than single product with the hope that when the returns from one enterprise is low, it is compensated by the higher returns from the other enterprise.
- Idle resources can be put to use.
- The enterprises having less income variability should be selected.
- Income variability for the enterprises would be lessened.
- Risk programming models and game theory models should be used to formulate whole farm plans under different risky situations.
- Stable enterprises should be preferred rather than risky enterprises particularly in situation of farm risks.

10.5.2. Insurance

- The farmers have many types of insurances to reduce production risk and financial risk.
- Crop insurance scheme reduces production risk.
- Livestock insurance provides safety against the fatal diseases of cattle.
- Farm assets can be insured against theft, fire or any other damage.

10.5.3. Agronomic Practices

- To reduce production risk, crop rotations, suitable varieties, deep tillage, mulching, *etc.,* should be adopted.

10.5.4. Market Risk Management

Selling the Farm Products at Different Points of Time

- Due to financial obligations, farmers sell their produce immediately after the harvest. During the harvesting periods generally prices will be low due to large arrivals. If the farmers avoid such selling practices, they can get remunerative prices for their products.
- The knowledge regarding supply and demand for the farm products and prices prevailing in different markets and other market information is essential.
- For perishable commodities, storage facilities, freezing facilities, processing facilities, *etc.,* are required to get remunerative prices.
- Hedging is another measure adopted by the farmers and the traders to safeguard against price-risk. Hedging practice is devised and followed by studying the future markets.

Government Price Policies and Programmes

- Minimum support price, procurement price, levy price, issue price, *etc.,* are set for various agricultural commodities by the Government every year to bring about price stabilization.
- Support price provides protection for the farmers against fall in the prices of farm products.

10.5.5. Financial Risk Management

- For reducing the financial risk we require many strategies, which are aimed at liquidity and solvency of the farm business. Solvency by definition refers to business ability to meet the long-term financial requirements of the farm. Liquidity strategies should aim at as to how to build up the farm business to meet the short-term cash requirements.
- At macro level fiscal policies formulated by RBI are aimed at to provide different measures to safeguard against the financial risk.

10.6. Crop Insurance

The Crop Insurance Scheme launched during February 2016 is in line with One Nation – One Scheme theme. It incorporates the best features of all previous schemes and at the same time, all previous shortcomings/weaknesses have been removed.

The highlights of this scheme are as under:

- There will be a uniform premium of only 2% to be paid by farmers for all Kharif crops and 1.5% for all Rabi crops. In case of annual commercial and horticultural crops, the premium to be paid by farmers will be only 5%. The premium rates to be paid by farmers are very low and balance premium will be paid by the Government to provide full insured amount to the farmers against crop loss on account of natural calamities.
- There is no upper limit on Government subsidy. Even if balance premium is 90%, it will be borne by the Government.
- Earlier, there was a provision of capping the premium rate which resulted in low claims being paid to farmers. This capping was done to limit Government outgo on the premium subsidy. This capping has now been removed and farmers will get claim against full sum insured without any reduction.
- The use of technology will be encouraged to a great extent. Smart phones will be used to capture and upload data of crop cutting to reduce the delays in claim payment to farmers. Remote sensing will be used to reduce the number of crop cutting experiments.
- The scheme was revamped in 2020 enabling voluntary participation of the farmers. It also made it convenient for the farmer to report crop loss within 72 hours of the occurrence of any event - through Crop Insurance App, CSC Centre or the nearest agriculture officer, with claim benefit transferred electronically into the bank accounts of the eligible farmer

10.6.1. National Crop Insurance Programme

NCIP is a Central Sector Scheme formulated by merging National Agricultural Insurance Scheme (NAIS), Pilot Weather Based Crop Insurance Scheme (WBCIS), Pilot Modified National Agricultural Insurance Scheme (MNAIS) and and Pilot Coconut Palm Insurance Scheme (CPIS). The scheme came into force from Rabi 2013-14 and covers all districts.

Loanee farmers will be covered on compulsory basis under the component-scheme of NCIP notified by the concerned state, while non-loanee farmers will choose either MNAIS or WBCIS component.

Private sector Insurers with adequate infrastructure and experience will be permitted to implement NCIP besides Agriculture Insurance company of India (AIC).

10.6.2. Insurance Products from Agricultural Insurance Company of India

Agriculture Insurance Company of India Limited (AIC) was incorporated under the Indian Companies Act 1956 on 20th December 2002 with an authorized share capital of INR 15 billion and paid up capital of INR 2 billion. AIC commenced business from 1st April, 2003.

AIC has taken over the implementation of National Agricultural Insurance Scheme (NAIS) which, until FY 2002-03 was implemented by General Insurance Corporation of India. In addition, AIC also transacts other insurance businesses directly or indirectly concerning agriculture and its allied activities.

10.7. Livestock Insurance

The Livestock Insurance Scheme, a centrally sponsored scheme was implemented on a pilot basis during 2005-06 and 2006-07 of the 10th Five Year Plan and 2007-08 of the 11th Five Year Plan in 100 selected districts. The scheme was later implemented on a regular basis from 2008-09 in 100 newly selected districts of the country.

The scheme was later subsumed as a component titled Risk Management and Insurance under the sub-mission on livestock development of National Livestock Mission. The component aims at management of risk and uncertainties by providing protection mechanism to the farmers against any eventual loss of their animals due to death and to demonstrate the benefit of the insurance of livestock to the people.

Coverage

The scheme is implemented in all the districts of the Country from 21.05.2014

Animal Covered

The indigenous / crossbred milch animals, pack animals (Horses, Donkey, Mules, Camels, Ponies and Cattle/Buffalo Male) , and Other Livestock (Goat, Sheep, Pigs, Rabbit, Yak and Mithun etc.) are covered under the purview of this component.

Central assistance

Benefit of subsidy is to be restricted to 5 animals per beneficiary per household for all animals except sheep, goat, pig and rabbit. In case of sheep, goat, pig and rabbit the benefit of subsidy is to be restricted based on "Cattle Unit" and one cattle unit is equal to 10 animals i.e a total of 50 animals. If a beneficiary has less than 5 animals / 1 Cattle Unit, s/he can also avail the benefit of subsidy.

Component	Pattern of assistance
Premium rates Premium rates for one year policy in • Normal Areas - 3.0% NER / Hill areas / LWE • affected areas -3.5%, Difficult areas - 4.0 %	**Normal areas** Central share 25%, State share 25% and Beneficiary share 50% for APL, and Central share 40%, State share 30%, and Beneficiary share 30% for BPL / SC / ST
Premium rates for three year policy in Normal Areas - • 7.5%, NER / Hill areas / LWE • affected areas - 9.0% • Difficult areas - 10.5 %	**NER / Hill areas / LWE affected areas** Central share 35%, State share 25% and Beneficiary sh are 40% for APL, and Central share 50%, State share 30%, and Beneficiary share 20% for BPL / SC / ST
	Difficult Areas Central share 45%, State share 25% and Beneficiary share 30% for APL, and Central share 60%, State share 30%, and Beneficiary share 10% for BPL / SC / ST

Process

An animal will be insured for its current market price. The market price of the animal to be insured w ill be assessed jointly by the beneficiary and the insurance company preferably in the presence of the Veterinary officer or the BDO. The minimum value of animal should be assessed by taking Rs.3000 per liter per day yield of milk or as per the price prevailing in the local market (declared by Government) for cow and Rs.4000 per liter per day yield of milk or as prevailing in the local market (declared by Government) for buffalo. The market price of pack animals (Horses, Donkey, Mules, Camels, Ponies and Cattle/Buff. Male) and Other livestock (Goat, Sheep, Pigs, Rabbit, Yak and Mithun) are to be assessed by negotiation jointly by owner of animal and by insurance company in the presence of veterinarians Doctor. In case of dispute the price fixation would be settled by the Gram Panchayat / BDO.

The animal insured will have to be properly and unique ly identified at the time of insurance claim. The ear tagging should, therefore, be full proof as far as possible. The traditional method of ear tagging or the recent technology of fixing microchips could be used at the time of taking the policy. The cost of fixing the identification mark will be borne by the Insurance Companies and responsibility of its maintenance will lie on the concerned beneficiaries. The nature and quality of tagging materials will be mutually agreed by the beneficiaries and the Insurance Company. The Veterinary Practitioners may guide the beneficiaries about the need and importance of the tags fixed for settlement of their claim so that they take proper care for maintenance of the tags. The tag already available on animal may be utilized with unique identity number subject to the condition that it is mutually agreed by farmer and agency and there shall not be any dispute in settlement of claims on account of utilization of existing tag.

While processing an insurance proposal, one photograph of the animal with the Owner and one photograph of the animal clearly with the EAR TAG visible shall be taken at the time of processing the insurance documentation. In case of sale of the animal or otherwise transfer of animal from one owner to other, before expiry of the Insurance Policy, the authority of beneficiary for the remaining period of policy will have to be transferred to the new owner.

Only four documents would be required by insurance companies for settling the claims viz. intimation with the Insurance Company, Insurance Policy paper, Claim Form and Postmortem Report. In case of claim becoming due, the payment of insured amount should be made within 15 days positively after submission of requisite documents. If an Insurance company fails to settle the claim within 15 days of submission of documents, the insurance company will be liable to pay, a penalty of 12% compound interest per annum to the beneficiary

10.8 Weather- Based Crop Insurance

Weather Based Crop Insurance aims to mitigate the hardship of the insured farmers against the likelihood of financial loss on account of anticipated crop loss resulting from incidence of adverse conditions of weather parameters like rainfall, temperature, frost, humidity etc.

Crops Covered

- Major Food crops (Cereals, Millets & Pulses) & Oilseeds
- Commercial / Horticultural crops

Farmers Covered

All farmers including sharecroppers and tenant farmers growing the notified crops in the notified areas are eligible for coverage. However, farmers should have insurable interest on the insured crop. The non-loanee farmers are required to submit necessary documentary evidence of land records and / or applicable contract / agreements details (in case of sharecroppers / tenant farmers).

All farmers availing Seasonal Agricultural Operations (SAO) loans from Financial Institutions (i.e. loanee farmers) for the crop(s) notified are covered on compulsory basis.

The Scheme is optional for the non-loanee farmers. They can choose between WBCIS and PMFBY, and also the insurance company.

Perils Covered

Following major weather perils, which are deemed to cause "Adverse Weather Incidence", leading to crop loss, shall be covered under the scheme.

- Rainfall – Deficit Rainfall, Excess rainfall, Unseasonal Rainfall, Rainy days, Dry-spell, Dry days
- Relative Humidity
- Temperature – High temperature (heat), Low temperature
- Wind Speed
- A combination of the above
- Hailstorms, cloud-burst may also be covered as Add-on/Index-Plus products for those farmers who have already taken normal coverage under WBCIS.

The perils listed above are only indicative and not exhaustive, any addition deletion may be considered by insurance companies based on availability of relevant data.

Risk Period

Risk period would ideally be from sowing period to maturity of the crop. Risk period depending on the duration of the crop and weather parameters chosen, could vary with individual crop and reference unit area and would be notified by SLCCCI before the commencement of risk period.

Premium Rates

The revised premium rates payable by the cultivator for different crops are as follows:

S.No.	Crops	Maximum Insurance charges payable by farmer (% of Sum Insured)
i)	Season - Kharif - Food & Oilseeds crops (all cereals, millets, & oilseeds, pulses)	2.0% of SI or Actuarial rate, whichever is less
ii)	Season - Rabi - Food & Oilseeds crops (all cereals, millets, & oilseeds, pulses)	1.5% of SI or Actuarial rate, whichever is less
iii)	Season - Rabi and Kharif - Annual Commercial / Annual Horticultural crops	5% of SI or Actuarial rate, whichever is less

The 'net premium payable in case of the insured loanee cultivator is financed by the Lending Bank.

Insurance Company Participating in WBCIS

The public sector and private sector General Insurance Companies empanelled by the Department of Agriculture & Cooperation (DAC) and Farmers Welfare, Government of India and selected by concerned State Government / Union Territory (UT) implement WBCIS.

FAQs on WBCIS

Weather based Crop Insurance Scheme (WBCIS) is a unique Weather based Insurance Product designed to provide insurance protection against losses in crop yield resulting from adverse weather incidences. It provides payout against adverse rainfall incidence (both deficit & excess) during Kharif and adverse incidence in weather parameters like frost, heat, relative humidity, un-seasonal rainfall etc. during Rabi. **It is not Yield guarantee insurance.**

Comparison between National Agricultural Insurance Scheme (NAIS) and Weather Based Crop Insurance Scheme (WBCIS)

S. No.	National Agricultural Insurance Scheme (NAIS)	Weather Based Crop Insurance Scheme (WBCIS)
1	Practically all risks covered (drought, excess rainfall, flood, hail, pest infestation,etc.)	Parametric weather related risks like rainfall, frost, heat (temperature),humidity etc.) are covered. However, these parametric weather parameters appear to account for majority of crop losses
2	Easy-to-design if historical yield data upto 10 years' is available	Technical challenges in designing weather indices and also correlating weather indices with yield losses. Needs upto 25 years' historical weather data
3	High basis risk [difference between the yield of the Area (Block / Tehsil) and the individual farmers]	Basis risk with regard to weather could be high for rainfall and moderate for others like frost, heat, humidity etc.
4	Objectivity and transparency is relatively less	Objectivity and transparency is relatively high
5	Quality losses are beyond consideration	Quality losses to some extent gets reflected through weather index
6	High loss assessment costs (crop cutting experiments)	No loss assessment costs
7	Delays in claims settlement	Faster claims settlement
8	Government's financial liabilities are open ended, as it supports the claims subsidy	Government's financial liabilities could be budgeted up-front and close ended, as it supports the premium subsidy.

How does Weather Based Crop Insurance Scheme (WBCIS) operate?

Weather based Crop Insurance Scheme (WBCIS) operates on the concept of "Area Approach" i.e., for the purposes of compensation, a 'Reference Unit Area (RUA)' shall be deemed to be a homogeneous unit of Insurance. This RUA shall be notified before the commencement of the season by the State Government and all the insured cultivators of a particular insured crop in that Area will be deemed to be on par in the assessment of claims. Each RUA is linked to a Reference Weather Station (RWS), on the basis of which current weather data and the claims would be processed. Adverse Weather Incidences, if any during the current season would entitle the insured a payout, subject to the weather triggers defined in the 'Payout Structure' and the terms & conditions of the Scheme.

The "Area Approach" is as opposed to "Individual Approach", where claim assessment is made for every individual insured farmer who has suffered a loss.

Weather based Crop Insurance Scheme (WBCIS) provides protection to the insured cultivators in the event of loss in crops yields resulting from the adverse weather incsidences, like un-seasonal/excess rainfall, heat (temperature), frost, relative humidity etc. Triggers are broadly fixed so as to capture the adverse incidence of weather parameters on crop yield.

Claims arise when there is a certain adverse deviation in Actual Weather Parameter Incidence in RUA (as per the weather data measured at RWS), i.e.,e.g. its "Actual temperature" within the time period specified in the Benefit Table is either less or more compared to the specified " temperature Trigger", leading to crop losses. In such case, subject to the terms and conditions of the Scheme, all insured cultivators under a particular crop shall be deemed to have suffered the same "adverse deviation" in temperature and become eligible for claims.

Weather experience (rainfall in particular) on a particular day could be different even in smaller geographical area, but, in a span of a fortnight, month or season it evens out. RWS at Block / Tehsil level, by and large, reflect the weather experience of individual cultivators within a RUA.

11

Natural Resource Economics

Abstract

This chapter examines natural resource economics, a multidisciplinary field that studies how economic and ecological systems interact to allocate natural resources on Earth. These resources, which include both human and nonhuman factors of production, are essential for the provision of goods and services. The chapter also explores the challenges of valuing and managing natural resources without market prices, and the implications for the sustainability of the food supply.

11.1. Introduction

Natural resource economics deals with the supply, demand and allocation of the earth natural resource. Main objective of natural resource economics is to better understand the role of natural resources in the economy in order to develop more sustainable methods of managing those resources to ensure their availability to future generations. An economy is a system of institutions and organizations that either help facilitate or are directly involved in the production and distribution of goods and services. Economic resources are the inputs we use to produce and distribute goods and services. The precise proportion of each factor of production will vary from product to product and from service to service, and the goal is to make the most effective use of the resources that maximizes output at the least possible cost. Misallocation or improper use of resources may cause businesses, and even entire economies, to fail. Resource economists study interactions between economic and natural systems, with the goal of developing a sustainable and efficient economy. It is a multi-disciplinary field of academic research. Economic resources are the factors used in producing goods or providing services. In other words, they are the inputs that are used to create things or help you provide services. Economic resources can be divided into human resources, such as labour and management, and nonhuman resources, such as land, capital goods, financial resources, and technology. The long-term viability of the food supply depends upon the sustainable use of natural resources. Unlike most agricultural inputs, such as fertilizer or animal feed, most natural resources do not have prices determined in the market. The field of Natural Resource Economics seeks to value natural resources to aid in the optimization of the production of goods and services from agricultural lands while protecting the environment.

11.2. Definition

Natural Resource Economics is application of economics to manage naturally occurring resources for human needs/wants with efficiency as the primary goal.

11.3. Natural Resources Goals and Objective Goal

Goals 1

- To protect and sustainably manage the County's natural resources.

Objectives

- Ensure that environmental and aesthetic qualities are considered in the planning process.
- Encourage wildlife management to reduce overpopulations of species. Encourage the sustainable management of the County's forest resources.
- Preserve and maintain environmentally sensitive areas.
- Retain and improve open space resources that provide significant ecological and environmental benefits.

Goals 2

- To ensure that development that does occur is sensitive to the environment.

Objectives

- To promote urban development which is environmentally sound and compatible with the natural resource base.
- Encourage that open space considerations be designed into development projects.

11.4. Scope of Natural Resource Economics

- Natural resource economics is the study of environments from which resources used in economic activity are extracted. Examples include agricultural production, exhaustible resource mineral extraction, renewable resource extraction such as forestry and fishing.
- Natural resource economics also includes the amenity value of wildlife, as in parkland and recreational forests.
- All natural resource decisions are governed by the legal environment in which such resources are produced and consumed.

- In turn, the legal environment, and the technology by which natural resources are consumed, also shapes the pricing of natural resource use.
- Beyond the legal environment and the state of technology that affect the level of prices, natural resource decisions also are governed by the underlying laws of thermodynamics.
- The first law states the constancy of matter and energy in the universe, a principle first enunciated by Lord Thompson in the late 18th century.
- The second law defines the efficiency of any energy conversion process: $B\text{-}1\text{-}T_1/T_2$, where T_1 is the ambient temperature and T_2 is the combustion temperature of any engine.
- Together these two laws also define the concept of entropy, which is a measure of disorder in any given environment. Typically, for any given economic process, one looks for low entropy choices as they appear to have the lowest transactions costs, after which one moves to higher entropy forms of energy consumption.
- Fossil and nuclear fuels tend to have relatively low levels of entropy, while renewable resources such as wind and solar, tend to have higher levels. Unless offset by technological innovations, higher entropy levels generally have deterred more widespread commercial applications.
- Market prices may reflect an imperfect guide to the economic value of natural resources, in which case, corrective, or shadow, prices, may be needed to achieve a given level of efficiency.
- Determining the economic value of natural resources requires not only an understanding of the laws of thermodynamics, but also how market forces achieve a given level of efficiency.
- Economic efficiency has a fairly precise set of norms in economic theory, but for any given degree of economic efficiency, it may not correspond to some level of environmental efficiency.
- Environmental efficiency refers to the ability of natural environments to make the most productive use of natural processes. From this we have the notion of the carrying capacity of an environment, and the notion of environmental sustainability.
- The challenge for economics is to find pricing solutions to natural resources that are consistent with environmental sustainability.
- Such pricing solutions typically arise in the context of economic models However, traditional economic models often overlook how the consequences of economic activity bear directly on the environment.
- Natural resource models inevitably raise normative questions regarding the present and the future, in particular the choice of an appropriate rate of

discount for a given level of investment in which the natural environment is affected.

- Beyond environmental and economic models is the larger question of global climate change the extent to which is taking place, the extent to which it is affected by human intervention, and whether global climate change is irreversible or part of some larger geophysical cycle.
- Recent studies such as the IPCC's Stern Report, place emphasis on the role of irreversibility on climate change, and on the normative value of undertaking action now to avoid irreversible losses in the future.
- The foundation of IPCC studies builds on climate change observations on which forecasting models have been derived. What many such forecasting models have ignored is the role of pricing, in particular the pricing of energy and natural resources in shaping any particular scenario.
- At a macroeconomic level, policy choices involving the environment require the continued accumulation of natural scientific evidence and its integration with economic forecasting models that can provide a consistent basis from which coherent public policies can be derived.
- Economists have relied on a variety of welfare policy criteria to evaluate any particular approach to natural resources and the environment. These criteria include: the Pareto criterion, the Kaldor criterion, the Samuelson criterion, and the Rawlsian criterion, among others.

11.5. Subject Matter

Economists research commercial and recreational resource usage and exploitation. Traditionally, natural resource economics concentrated on fishery, forestry, and mineral models. However, many other concerns, including as air, water, and global climate, have become increasingly relevant in recent years. Natural resource economics is studied academically, and the conclusions are utilised to shape and influence environmental policy.

Examples of areas of study in natural resource economics include:

- Welfare theory
- Pollution control
- Resource exhaustibility
- Environmental management
- Resource extraction
- Non-market valuation
- Environmental policy.

Natural resource economists can also do research on issues such the effects of agriculture, transportation, and urbanization on the environment, the use of land in developing and developed nations, the environment and trade, and climate change.

11.6. Importance of Natural Resource Economics

Successful stewardship of the land entails understanding the complex interrelationships between environmental and market forces. Research is pursued to understand why producers adopt sustainable technologies and identify incentives as well as costs and benefits to producers and the environment. Resource economists provide science-based information to help agriculture producers balance production demands with elements central to agricultural sustainability, including.

- Maintenance and improvement of air and soil quality.
- Adaptation and management of climate change dynamics
- Efficient use of surface and groundwater supplies
- Maintenance of landscape vistas, wild plant and animal habitats, and recreational areas

Natural resource economics also demonstrates how policy incentives might motivate better choices by agriculture producers and makes predictions about the potential side effects of those choices.

11.7. Environmental and Ecological Economics

The insights into sustainability provided by mainstream economics are taken much further by environmental and ecological economists. The main areas of contribution include the following:

- A classification of sustainability views according to assumptions about the conservation of natural resources
- Extending the analysis of externalities to provide a basis for designing antipollution policies and deciding on the resources it is desirable to devote to avoiding pollution
- A range of methodologies for evaluating the services provided by environmental assets and social capital to extend the inclusiveness of Cost Benefit Analysis.
- Models for projecting the pricing and depletion of finite resources.
- Assessments of the implications of various access regimes governing the harvesting of renewable resources.

Ecological economics and environmental economics cover a lot of the same ground. The most important variation is in the orientation. Incorporating environmental resources and services is a goal of environmental economics, which often adopts the Neo-classical paradigm as an examination of the economic system. The health of intricately interconnected ecological systems is given top emphasis in ecological economics, which also looks at how economic behaviour may be changed to achieve that goal.

11.8. Natural Resources Classification

1. **Natural resources are often classified into renewable and non-renewable resources.**

a) **Renewable resources:** A natural resource that may refill over time, whether through biological reproduction or other naturally occurring cycles, is said to be renewable. Due to their ability to regenerate or never run out, they are referred to as renewable. Natural resources that can be replenished and used repeatedly in our lifetimes are known as renewable resources. Fresh water, wood, oxygen, and sun energy are a few examples. The natural environment of the Earth and the main elements of its ecosphere are renewable resources. A resource's viability can be determined in large part by the results of its life cycle assessment. Renewable energy sources could come from renewable resources. In order to exploit renewable resources sustainably, one must maintain a positive renewal rate, which can lessen land degradation, soil deterioration, air pollution, and habitat damage.

i. **Organic renewable resource:** When renewable resources come from living things, they are considered organic, such as manure.

ii. **Inorganic renewable resource:** These are nonliving resources like solar power, wind energy and hydroelectric energy.

b) **Non-renewable natural resources:** Natural resources that cannot be replaced during our lifetimes are referred to as non-renewable resources. They consist of earth minerals, fossil fuels, metal ores, and in some cases, groundwater. A non-renewable resource, usually referred to as a finite resource, is one that cannot be sustainably extracted economically over a considerable period of time. Organic fuel with a carbon base is one illustration. With the use of heat and pressure, the intestinal organic material transforms into a fuel like gas or oil. All non-renewable resources include fossil fuels (such as coal, petroleum, and natural gas) and some aquifers.

Carrying capacity: Use of natural resource services is compared with defined bio- physical limits for the supply of such services.

2. Resources can also be classified on the basis of their origin i.e. biotic and abiotic

- **Biotic resources:** Biotic resources are derived from animals and plants (Le-the living world). Biotic is a living component of a community, for example organisms, such as plants and animals.
- **Abiotic resources:** Abiotic resources are derived from the non-living world eg., land, water, and air. Mineral and power resources are also abiotic resources some are derived from nature. In biology and ecology, abiotic components are non-living chemical and physical factors in the environment which affect ecosystems

Natural resources are also categorized based on the stage of development

- Potential Resources are ones that are already known to exist and could be applied later. For instance, sedimentary rocks in many areas of Kuwait and India may contain petroleum, but this is still only a prospective resource until it is actually discovered and used.
- The resources that have been examined, their quantity and quality established, and are currently being used are the actual resources. The Mumbai High Fields, for instance, are extensively used to obtain petroleum and natural gas. A reserve resource is the portion of an actual resource that can be developed profitably using currently available technology, whereas a stock resource is the portion of an actual resource that cannot be developed profitably due to a lack of technology.

11.9. Characteristics of Natural Resources

Resources have three main characteristics namely

1. Utility: In economics, utility is a measure of satisfaction, referring to the total satisfaction received by a consumer from consuming a good or service
2. Scarcity: Scarcity is the fundamental economic problem of having humans who have unlimited wants and needs in a world of limited resources. It states that society has insufficient productive resources to fulfill all human wants and needs.
3. Resource depletion: Resource depletion is an economic term referring to the exhaustion of raw materials within a region. Resource depletion is most commonly used in reference to farming, fishing, mining, and fossil fuels.

Causes of resource depletion of natural resources

- Over-consumption/excessive or unnecessary use of resources
- Non-equitable distribution of resources
- Overpopulation
- Slash and burn agricultural practices, currently occurring in many developing countries
- Technological and industrial development
- Erosion
- Irrigation
- Mining for oil and minerals
- Aquifer depletion
- Forestry
- Pollution or contamination of resources

11.10. Economic Approaches to Resource Management

In economic approaches to resource management, the common denominator is often some type of assessment that may be connected to personal welfare. The majority of natural resource and environmental issues can be examined using a comprehensive framework provided by economics. Optimal resource extraction and utilisation, in particular as examined by the Hotelling's rule. Economic sustainability indicators are derived from the dubious sustainability tenet that the complete capital stock must be preserved. Several simplifying presumptions form the foundation of the fundamental Hotelling Rule. It is assumed that all market participants have complete knowledge and that the total stock of resources is known and of equal quality. Non-renewable resource management is primarily concerned with how a resource pool should be exploited, rather than with sustainability.

11.11. Difference Between NRE and Agricultural Economics

Natural Resource economics is study of ways and means in achieving the optimum utilization of natural resources, both renewable and non-renewable.

Agricultural Economics is an applied science wherein principles of choice are applied to the field of farming/agriculture for maximizing the benefits.

	NRE	Agricultural Economcs
Subject of Interest	Economics study of resource	Economics study of input use and output
Focus upon	Optimum extraction of resource	Optimum dose of input use and optimum level of output to produce.
Condition for optimality	Price = Marginal Cost + Opportunity cost of resources not extracted (Rent)	Marginal Cost (MC) is equal Marginal Revenue (MR)
Cost concepts used	User cost = royalty = Scarcity rent (opportunity cost)	Marginal cost
Analysis	Dynamic analysis (use of differential and difference equation) Dynamic analysis (use of differential and difference equation)	Static analysis (use of static differential calculus)
Role of markets	Well-defined markets may/ may not exist. For example, water markets, land markets and market for recreation. Since total economic valuation is important and markets does not provide full valuation of resource and therefore, valuation of resource is more relevant than price .	Well-defined markets exist. For example, commodity markets, agricultural input market. Hence, market price is relevant.
Stakeholders	Resource extractor, user, processor, Resource owner and Government	Farmers and Farm Families, Traders, Processors, Consumers and Government
Policy framework	Long term macro-economic policy is must. For example, National Water policy, National Mining Policy.	Relatively short term policies. For example, fertilizer policy.

11.12. Externalities

The allocation of resources to productive uses results from consumers and producers making decisions with the aim of maximizing satisfaction and profits, respectively. Private costs and benefits are taken into account in deciding purchases and organizing production. Social costs and benefits include the costs and benefits of consumer and producer but also costs borne by those who are not participating in that particular market. These are known as external costs and benefits or externalities. In economics, an externality is the cost or benefit that affects a party who did not choose to incur that cost or benefit. Externalities often occur when a product or service's price equilibrium cannot reflect the true costs and benefits of that product or service. This causes the externality competitive equilibrium to not be optimality. External costs and benefits can arise through both consumption and production. Externalities create divergence between social benefit and private benefit and between social cost and private cost.

It is of two types: Positive and Negative externality

Negative externality occurs when one agent's action leads to unintended loss to other economic agent without paying compensation to the latter. For example, pollution from factory occurs because of production of goods. However, the intention of the factory owner/manger was not to harm the environment. Thus, an unintended loss occurs to environment and all creatures living in that environment. But this loss is not compensated leading to negative externality. These unintended costs imposed on other economic agent is called as external cost.

Another classic example of negative externality is when farmers use fertilizer. Usage of fertilizer leads to benefits to farmers by enhancing yields on farm but leads to loss to *eutrophication* of water bodies by run-offs from fields where fertilizer was applied. This causes loss to a fisherman whose fish production is adversely affected. However, it is difficult to find out as to from which field run-offs were there and nearly impossible to impose cost on the farmer from whose field such run-offs were there. Thus, neither the farmer had any incentive for causing eutrophication and consequent loss to fisherman but also farmer is not paying any loss incurred to fisherman. Thus, negative externality is said to be present in this case.

Positive externality occurs when one agent's action leads to unintended benefits to other economic agent without payment of compensation to the former. For example, honey bee while collecting pollens does pollination which benefits farmer but farmer does not pay the compensation to honeybee for performing pollination. Thus, actions of honeybee leads to positive externality. These unintended benefits are called as external benefits.

In market, producer receives full payment for their produce and consumer pays full cost of the produce. Thus, private benefits equals private cost. Presence of externality leads to market failure because externality takes away pricing power of the market.

11.13. Inefficiency and Welfare Loss Due to Externality and Solution Thereof

In the **positive externality** case, currently the producer is producing 'Q' quantity of goods at point where private marginal benefit (PMB) is equal to private marginal cost (PMC) [PMB=PMC]. The price per unit equals the cost per unit effectively meaning that benefit equal to cost. But this production process has positive effect on environment which leads to some external benefits which are enjoyed by the society. Now, if social marginal cost (SMC) is equal to private marginal cost [PMC=SMC] and social marginal benefits higher (SMB) than private marginal benefits (PMB) at each level of production then environmentally or socially optimum quantity is at 'Q1'where SMB=SMC. Thus, it is evident that due to presence of positive externality, produce is producing less quantity than is environmentally/socially desired. Thus, presence of positive externality leads to inefficiency. The solution to this problem is to provide subsidy per unit of production equal to 'AB'to the producer so much that private marginal benefit equals social marginal benefits. This subsidy is referred to as Pigovian Subsidy named after famous Welfare Economist, Arthur Cecil Pigou who gave this concept. However, providing this subsidy leads to welfare loss equal to area of triangle 'ABC'.

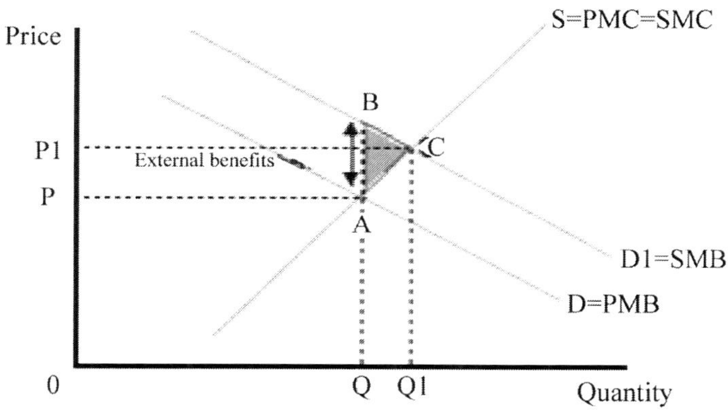

A case of Positive Externality

In **negative externality** case, lets assume that the production of goods involves pollution which exerts external cost to the society in terms of harm to the environment. Currently the producer is producing the 'Q' quantity of goods where private marginal benefit is equal to private marginal cost [PMB=PMC]. Let's assume that private marginal benefits are same as social marginal benefits [PMB=SMB]. According to social/environmental optimum, the producer should reduce the output to the point where SMB=SMC, i.e., to 'Q1'. Therefore, in case of negative externality, the producer is producing more than socially desired quantity. Hence over production is taking place. Thus, presence of negative externality has introduced inefficiency in the production system. In order to reduce the excess of external cost on society, the Government can introduce tax on the production of this good. Such a tax is referred to as Pigovian Tax after famous Welfare Economist, Arthur Cecil Pigou. Let's say his tax per unit of production is equal to marginal external cost. Imposition of such tax lead stoequality of SMC and SMB. Thus, after introduction of Pigoviantax, producer would be producing socially desired quantity of goods.

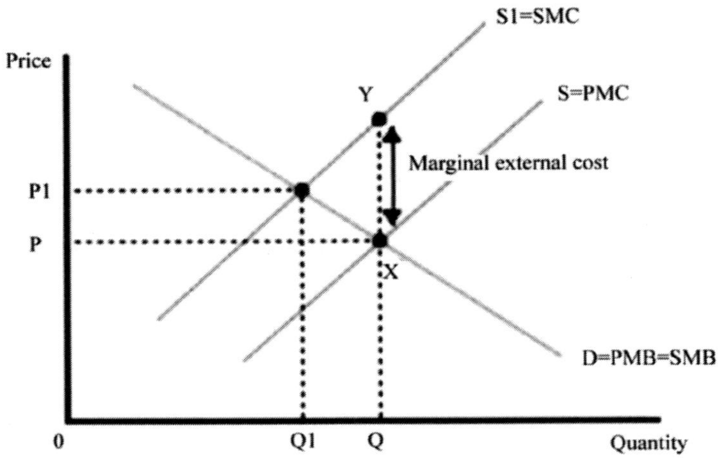

A case of Negative Externalities

11.14. Market and Non Market Valuation of Natural Resource

The economic value of natural resources can be estimated by the price individuals will pay in order to obtain the benefits of the resources. The non-market values of that communities attach to natural resources, such as recreational, existence and protection values. To support efficient decision making, environmental valuation techniques are needed to quantify the value impacts resulting from changed catchment management. Several approaches to environmental valuation are there i.e. revealed preference techniques include travel cost and hedonic pricing methods, stated preference techniques include contingent valuation and choice experiments. Stated preference techniques are increasingly being used to estimate non-market values. Nonmarket valuation is a method to estimate the value of goods and services that are not commonly bought and sold in markets.

Non-market valuation is a sub-field within environmental economics that deals with theoretical and practical aspects of estimating monetary values on non-marketed environmental goods and services. Specifically, techniques include the travel cost method of recreational demand, the hedonic property pricing model, averting behavior analysis, contingent valuation, contingent rating/ ranking, and stated choice experiments.

Benefit-cost analysis

Benefit-cost analysis is a systematic method that compares the accumulative social benefits with the opportunity costs. Benefit-cost analysis can be more difficult to apply when the benefits or cost do not have a market value.

Travel Cost Model

The travel cost model estimates the implicit price of natural resources based on outlays of time and travel expenses. An evaluation of these costs incurred in using a natural resource (e.g. a state park) can be used to estimate the regional demand for the resource. The aggregate value of the resource can then be inferred from a combination of this demand and the number of visits to the site over time. This model can be applied to many recreational activities that utilize the more intangible natural resources such as bird watching.

Random Utility Models

The random utility model is similar to the travel cost model except that rather than focusing on the number and overall costs of trips to different sites, it estimates the value individuals place on particular sites based on the attributes of the site.

Hedonic Pricing Methods

Hedonic pricing methods involve a comparison of specific sites that differ only by some environmental attribute such as proximity to a forested area or availability of a view. Comparison of the property value between two house sites provides an indication of the value to the individual paying a higher price for the site with the preferred attribute.

Contingent Valuation Method

The contingent valuation method is a survey- or questionnaire-based approach to estimating the value of non-market goods. This is the primary method used for placing value on resources that cannot be valued by the indirect methods. This is the only method that can be used to estimate "existence values".

Existence Value

This is the value that individuals place on natural resources that they want to remain unaltered, even though they may not use or visit the area.

Economic Value

It is one ways to measure the value of a resource. Economic values are useful to consider when making economic choices – tradeoffs in allocating resources. Measures of economic value are based on individual preferences

Valuation: Ecological and Environmental Economics

Habitat : Opportunity cost approach, Replacement cost approach, Land Value approach & Contingent Valuation Approach

Air and water quality: Cost effectiveness of prevention, Preventive expenditure, Replacement - reallocation costs, etc

Recreational: Travel Cost Approach and Contingent Valuation Approach.

Aesthetics, biodiversity, cultural, historical assets: Contingent Valuation Approach

Use Values (i.e., personal use): Use Values are divided into market and non-market commodities:

a. Market commodities are those which are traded in competitive cash markets. Private timber is a good natural resource example of a market commodity.

b. Non market commodities involve public goods, open access or unique resources potentially constituting natural monopoly situations.

c. option value: willingness to pay by a person to guarantee that a resource would be available should he/she choose to use it in the future.

2. Non Use Values

i. Bequest value: WTP by a person so that a resource might be passed on to future generations.

ii. Existence value: WTP by a person so that a resource will continue to exist today even if he/she never uses the resource. There is no personal use.

Common Methods for Estimating Prices for Publicly Provided Natural Resources

A. Market or Quasi-Market Methods of Price Estimation

1. **Competitive market transaction data** - Competitive price information from actual private market transactions as a basis for valuation. i.e. Private market annual timber sales data is good example of market transaction data.

2. **Residual valuation** - used in situations where private markets for natural resources may not be active.

3. **Change in net income** - The general idea behind this pricing method is to estimate what the natural resource contributes to producer's profits (net income),

B. Non-Market and Non-use Methods

1. Travel cost method (TCM)

It is most widely-used method for determining the demand WTP for outdoor recreation sites. It is a revealed preference method. TCM was first suggested by Hotelling and later Clawson, and has since been used extensively to estimate demand for recreation sites.

2. Hedonic TCM

Define Hedonic method - the Hedonic price method is used to estimate the value of an attribute (e.g., site quality) of a good (e.g., recreation site). In other words, the hedonic model, the multiple regressions, decomposes the total value of the good into the value of its several attributes.

3. Random Utility Models

RUMs are used to measure the economic impacts of changes in recreation site quality. It is a type of TCM. RUMs use the form of a multinomial logistical regression model (i.e., logit model) where the predicted dependent variable (i.e., probability of visit given values of independent variables) is restricted to lie between the unit interval, 0 to 1.

4. Contingent Valuation Method

Many important valuation problems are non-observable; i.e., no value measures can be derived from observing individual choices in a market. Some of these situations involve potential rather than actual policy changes. Such cases call for value measurement methods which use hypothetical, constructed markets.

11.15. Important Issues in Economics & Management of Common Property Resources of Land, Water, and Forest Resources Etc.

Natural Resource	Objectives	Issues / Problems
1. Land	1. Efficient use – estimation of return, value and rent of land. 2. Soil fertility, microbial life and impact on productivity.	1. Decline in the soil fertility, degradation etc. 2. Imperfection in and market, transaction
2. Water	1. Efficiency in water use and pricing. 2. Sharing – inter and intra generational equity	1. Depletion of Ground Water extraction exceeding the recharge. 2. Well interference & externality. 3. Emerging water markets.
3. Fishery	1. Optimum harvest and optimum rotation period. 2. Regulation of fishing- tax, quota and co-operative arrangement	1. Property issue – common property problems, harvesting under open access. 2. Extinction of species
4. Forestry	1. Optimum rotation and harvest of timber 2. Harvest and valuation of Non Timber Forest Products (NTFPs) 3. Valuation and pricing of environmental benefits i.e. positive externality. 4. Total economic valuation of benefit	1. Non renewability or depletion. 2. Property right issues – ex. Community forest. 3. Non market nature of environmental benefits
5. Mineral and oil	1. How much to be extracted in each period. 2. Pricing and costing. 3. Estimation and regulation of externalities.	1. External effect i.e., negative externalities-in extraction, use and afterwards. 2. Depletion and alternative source
6. Biodiversity	1. Valuation and the impact of sustainability. 2.Biodiversity & intergenerational equity.	1. Market failure, Policy failure. and developmental impact 2. Species loses.

A common problem faced with common property resources is Tragedy of commons. This concept was given by Garret Hardin in 1968. According to this concept, rationally acting individuals when acting independently according to their self-interest will deplete the common property resource in the long run. This happens when there is community tank where each person from community can catch fish without restriction. Then if each person starts to maximize their own catch of fish then gradually each person will be able to catch less fish with more efforts and finally one day all fish in the tank will be depleted. The solution to this problem is to have co-operatives to manage it.

Appendix

Choose the correct answer

1. Additional income obtained in a production programme from producing an additional unit of output is

   a) Marginal cost b) Marginal input

   c) Average revenue d) Marginal revenue

2. When Average Variable Cost is decreasing, it implies input use efficiency

   a) Increasing b) Constant

   c) Decreasing d) None

3. When APP is maximum, AVC is?

   a) Maximum b) Constant

   c) Minimum d) None

4. Additional returns received from using additional unit of input is known as

   a) MR b) MVP_X

   c) MIC d) MC

5. Which of the following represents unit cost of production i.e., the cost of producing one unit of output?

   a) AFC b) ATC

   c) AVC d) MC

6. Which of the following is a continuous falling curve?

   a) AFC b) AVC

   c) ATC d) MC

7. When the business activity is nil, total cost is equal to?

   a) Variable cost b) Fixed cost

   c) Marginal cost d) Zero

8. When the business activity is nil, variable cost is equal to?
   a) Opportunity b) Fixed cost
   c) Marginal d) Zero
9. Which of the following curves cuts the ATC curve at its lowest point?
   a) MC b) AVC
   c) TC d) AFC
10. Which of the following is correct?
    a) MC curve is a continuous falling curve
    b) AVC curve lies above ATC curve
    c) When MC curve is below ATC curve, ATC is falling
    d) When MC curve is above ATC curve, ATC is falling
11. In a production function, at the end of zone I, _________ is maximum
    a) TPP b) MPP
    c) APP d) None
12. In a production function, at the end of zone II, _________ is minimum
    a) TPP b) MPP
    c) APP d) None
13. In zone III of production function, _________ is minimum
    a) TPP b) MPP
    c) APP d) None
14. The level of fixed resources is high in _________ of production function
    a) Zone I b) Zone II
    c) Zone III d) None
15. The level of variable resources is high compared to fixed resources in _________ of production function
    a) Zone I b) Zone II
    c) Zone III d) None
16. The level of variable resources is excess in quantity in _________ of production function
    a) Zone I b) Zone II
    c) Zone III d) None
17. There is scope for reorganization of resources in _________ of production function
    a) Zone I b) Zone II
    c) Zone III d) Zones I and III

18. There is no scope for reorganization of resources in _________ of production function

a) Zone I | b) Zone II
c) Zone III | d) Zones I and III

19. Technical efficiency of both fixed and variable resources increases in _________ of production function

a) Zone I | b) Zone II
c) Zone III | d) Zones I and III

20. Technical efficiency of both fixed and variable resources decreases in _________ of production function

a) Zone I | b) Zone II
c) Zone III | d) Zones I and III

21. In a classical production function, the elasticity of production is greater than one, where

a) The point of equality between average physical product and marginal physical product
b) Upto the point where average physical product is maximum
c) The point where total physical product is maximum
d) The point where marginal physical product is maximum

22. The point, where the elasticity of production remains zero is lying at

a) AR = MR | b) MR > AR
c) AR > MR | d) MR = TR

23. Which of the following costs includes managerial services component

a) Cost Al | b) Cost C3
c) Cost B1 | d) Cost D

24. Crop Yield Index is an

a) Physical efficiency aggregate measure
b) Physical efficiency ratio measure
c) Financial efficiency aggregate measure
d) Financial efficiency ratio measure

25. At Break even point in a production programme, an Economist advise the farmer

a) To continue production programme temporarily
b) To stop production programme temporarily
c) To stop production programme permanently
d) To increase the fixed costs

26. The long-run average cost curve
    a) Passes through the minimum points of all short-run average total cost curves
    b) Passes through the minimum points of all short-run average variable cost curves.
    c) Is the envelope of all short-run average total cost curves.
    d) Us the envelope of all short-run average variable cost curves.
27. Due to depreciation, the value of an asset
    a) Decreases b) Increases
    c) Remains constant d) None
28. Which of the following is correct?
    a) MC curve is a continuous falling curve
    b) AVC curve lies above ATC curve
    c) When MC curve is above ATC curve, ATC is rising
    d) When MC curve is above ATC curve, ATC is falling
29. Which of the following is correct?
    a) MC curve is a continuous falling curve
    b) AVC curve lies above ATC curve
    c) When MC curve is below AVC curve, AVC is falling
    d) When MC curve is above AVC curve, AVC is falling
30. Which of the following is correct?
    a) MC curve is a continuous falling curve
    b) AVC curve lies above ATC curve
    c) When MC curve is above AVC curve, AVC is rising
    d) When MC curve is above AVC curve, AVC is falling
31. Which of the following is the correct representation of factor-factor relationship?
    a) Y = f(X1 X2/X3 X4 .....Xn) b) Y = f(X1 X2 X3/X4 .....Xn)
    c) Y = f(X1 X2 X3 X4 .....Xn) d) Y = f(X1/X2 X3 X4 ..... Xn)
32. Which of the following is not a good measure of the size of a farm business?
    a) Machinery inventory b) Number of work units
    c) Capital investment per worker d) Tillable acres

33. Isoquant is also known as
    a) Product indifference curve
    b) Production possibility curve
    c) Price line
    d) Iso resource curve
34. A number of isoquants on the same graph is called
    a) Product indifference curve
    b) Production possibility curve
    c) Iso resource curve
    d) Isoquant map / contour
35. Which of the following is true with reference to isoquant?
    a) It is convex to origin
    b) They do not intersect each other
    c) The slope indicates MRTS of two resources
    d) All the above
36. The products which when produced together results in the increase in the production of one of the products without affecting the other are called
    a) Complementary products
    b) Competitive products
    c) Joint products
    d) Supplementary products
37. The products which when produced together results in affecting the production of both the products are called
    a) Complementary products
    b) Antagonistic products
    c) Competitive products
    d) Supplementary products
38. For complementary enterprises, MRPS is
    a) Greater than zero
    b) Equal to zero
    c) Less than zero
    d) Infinity
39. For supplementary enterprises, MRPS is
    a) Greater than zero
    b) Equal to zero
    c) Less than zero
    d) Infinity
40. For competitive enterprises, MRPS is
    a) Greater than one
    b) Less than zero
    c) Equal to one
    d) Infinity
41. When two products substitute at increasing rate, it is better to go?
    a) Specialization
    b) Diversification
    c) Stop the production programme
    d) Both (a) and (b)

42. When two products substitute at constant rate, it is better to produce?

a) Specialization b) Diversification

c) Stop the production programme d) Both (a) and (b)

43. When two products substitute at decreasing rate, it is better to produce?

a) Specialization b) Diversification

c) Stop the production programme d) Both (a) and (b)

44. When two products substitute at increasing rate, it is better to produce?

a) One product only b) Two products

c) Stop the production programme d) None

45. When two products substitute at decreasing rate, it is better to produce?

a) One product only b) Two products

c) Stop the production programme d) None

46. Which of the following is correct?

a) Social cost = Private cost + External cost

b) Social cost = Private cost - External cost

c) Social cost = External cost - Private cost

d) Social cost = Private cost + Opportunity cost

47. $MRTS_{x1x2}$ is given by

a) $\Delta X2/\Delta X1$ b) $\Delta X2 \bullet \Delta X1$

c) $\Delta X1/\Delta X2$ d) $\Delta X2+\Delta X1$

48. Which of the following is the correct representation of least cost combination of resources

a) $(\Delta X_2/\Delta X_1) = (PX_1/PX_2)$ b) $(\Delta X_2/\Delta X_1) + (PX_2/PX_1)$

c) $(\Delta X_2/\Delta X_1) = (PX_2/PX_1)$ d) $(\Delta X_2/\Delta X_1) / (PX_2/PX_1)$

49. $MRTS_{x2x1}$ is given by

a) $\Delta X2/\Delta X1$ b) $\Delta X2 \bullet \Delta X1$

c) $\Delta X1/\Delta X2$ d) $\Delta X2 + \Delta X1$

50. The use of inputs which can be reshuffled in a production programme to produce same level of output are called

a) Complements b) Substitutes

c) Supplements d) None

51. Inputs which increase the output when combine in fixed proportion are called

a) Complements b) Substitutes

c) Supplements d) None

52. Family labour and hired labour, home produced grain and purchased grain are

a) Perfect Complements b) Perfect Substitutes

c) Perfect Supplements d) None

53. Tractor hours and man hours are

a) Perfect Complements b) Perfect Supplements

c) Perfect Substitutes d) None

54. Which of the following is a source of diseconomies of scale?

a) As a firm expands it finds itself depending more on hierarchical means of coordinating it's employees' activities

b) As a firm expands it's output labor begins to specialize

c) Individual employee incentives become clearer

d) Individual departments and divisions tend to work more as a team

55. ________ type of farming ensures better utilization of farm by-products

a) Diversified farming b) Specialized farming

c) Mixed farming d) Ranching

56. ________ type of farming ensures maintenance and improvement of soil fertility

a) Diversified farming b) Specialized farming

c) Mixed farming d) Ranching

57. ________ type of farming ensures balanced use of farm labour

a) Diversified farming b) Specialized farming

c) Mixed farming d) Ranching

58. ________ type of farming involves grazing and multiplication of livestock on natural surroundings

a) Diversified farming b) Mixed farming

c) Specialized farming d) Ranching

60. A ranch land is

a) Unfit for cultivation b) Fit for extensive cultivation

c) Fit for intensive cultivation d) Fit for prawn cultivation

61. Which of the following is a system of farming?
    a) Diversified farming
    b) Specialized farming
    c) Cooperative farming
    d) Ranching

62. Which of the following is a type of farming?
    a) Collective farming
    b) Peasant farming
    c) Cooperative farming
    d) Ranching

63. ________ is a system of farming under which all or part of the agricultural operations are carried out jointly by the farmers on a voluntary basis, but each farmer retains the right on his own land
    a) Collective farming
    b) Peasant farming
    c) Cooperative farming
    d) Capitalist farming

64. ________ is a system of farming under which both ownership and operationship are individual
    a) Cooperative better farming
    b) Cooperative joint farming
    c) Cooperative collective farming
    d) Cooperative tenant farming

65. ________ is a system of farming under which both ownership and operationship are collective
    a) Cooperative better farming
    b) Cooperative collective farming
    c) Cooperative joint farming
    d) Cooperative tenant farming

66. The farm management principle which will guide the farmer in selecting the least cost method of production of enterprises is
    a) Opportunity cost principle
    b) Principle of comparative advantage
    c) Principle of factor substitution
    d) Principle of product substitution

67. Which of the following is a power function?
    a) Linear function
    b) Cobb-Douglas function
    c) Quadratic function
    d) None

68. In Economics, size of farm refers to
    a) Number of acres
    b) Value of production
    c) Total labour employed
    d) Total capital employed

69. ________ refers to size of the farm, which keeps the farmer fully employed in the farm business and the income he derives is fully sufficient to sustain himself and his family

a) Minimum efficient size b) Economic size

c) Optimum size d) Family holding

70. ________ refers to size of the farm, which gives a reasonable standard of living to the farm family

a) Minimum efficient size b) Economic size

c) Optimum size d) Family holding

71. Better use of land, better marketing and better management are the features of

a) Diversified farming b) Specialized farming

c) Mixed farming d) Ranching

72. Which of the following is the social factor influences the selection of farm in an area?

a) Topography b) Transport facility

c) Local taxes d) Accessibility to school

73. Greater risk is involved in ________ type of farming

a) Diversified farming b) Specialized farming

c) Mixed farming d) Ranching

74. ________ type of farming is a combination of crop enterprise with a significant contribution from livestock raising

a) Diversified farming b) Mixed farming

c) Specialized farming d) Ranching

75. When two products substitute at constant rate, it is better to produce?

a) One product only b) Two products

c) Stop the production programme d) None

76. When two products have complementary relationship, it is better to produce?

a) One product only b) Two products

c) Stop the production programme d) None

77. When two products have competitive relationship, it is better to produce?

a) One product only b) Two products

c) Stop the production programme d) None

78. When two products have supplementary relationship, it is better to produce?

a) One product only b) Two products

c) Stop the production programme d) None

79. In constant returns to scale, the gap between successive isoquants with increase in output in the production programme

a) Remains same b) Increases

c) Decreases d) None

80. In decreasing returns to scale, the gap between successive isoquants with increase in output in the production programme

a) Remains same b) Increases

c) Decreases d) None

81. In increasing returns to scale, the gap between successive isoquants_ with increase in output in the production programme

a) Remains same b) Increases

c) Decreases d) None

82. The summation of elasticities of production in a Cobb-Douglas production function results in

a) Returns to scale b) Production function

c) MRTS d) MRPS

83. Which of the following is the correct representation of Returns to scale in Cobb-Douglas production function?

a) Σb_i b) $\Delta Y_2/\Delta Y_1$

c) $\Sigma a_i b_i$ d) $(\Delta Y_2/\Delta Y_1) = (PY_1/PY_2)$

84. Which of the following is the correct representation of optimum product combination

a) $(\Delta Y_2/\Delta Y_1) = (PY_1/PY_2)$ b) $(\Delta Y_2/\Delta Y_1) + (PY_2/PY_1)$

c) $(\Delta Y_2/\Delta Y_1) = (PY_2/PY_1)$ d) $(\Delta Y_2/\Delta Y_1)/(PY_2/PY_1)$

85. $MRPS_{Y1Y2}$ is given by

a) $\Delta Y2/\Delta Y1$ b) $\Delta Y1/\Delta Y2$

c) $\Delta Y2 \bullet \Delta Y1$ d) $\Delta Y2+\Delta Y1$

86. The products which are produced through single production process and one of the products cannot be produced alone but must be accompanied by the other product are called ________

a) Complementary products b) Substitution products

c) Joint products d) Supplementary products

87. In case of joint products, the production possibility curve is ________

a) 'L' shaped b) Inverted 'L' shaped

c) Concave d) 'U' shaped

88. In case of constant rate of substitution of products, the production possibility curve is

a) 'L' shaped b) Inverted 'L' shaped

c) Straight line d) 'U' shaped

89. In case of increasing rate of product substitution, the production possibility curve is

a) 'L' shaped b) Inverted 'L' shaped

c) Concave shaped d) 'U' shaped

90. In case of decreasing rate of product substitution, the production possibility curve is

a) Convex shaped b) Inverted 'L' shaped

c) Concave shaped d) 'U' shaped

91. The products which when produced together results in the increase in the production both the products are called

a) Complementary products b) Competitive products

c) Joint products d) Supplementary products

92. The products which when produced together results in the increase in the production of one of the products at the cost of the other are called

a) Complementary products b) Competitive products

c) Joint products d) Supplementary products

93. Uncertainty in agriculture can be minimized by the farmer through

a) Selecting HYV for crop cultivation

b) Specialization in crop production

c) Diversification of enterprises

d) Studying Agro-Meteorology subject

94. Which of the following is not a system of farming on the basis of ownership and operationship?

a) Dry farming b) State farming

c) Peasant farming d) Capitalistic farming

95. When total cost is divided by total quantity of production, it is called
   a) MC b) Opportunity cost
   c) AC d) Sunk cost

96. Each point on a production possibility frontier always represents ________
   a) Factors of production b) A shortage
   c) A trade-off d) Human capital

97. Email is an example of which type of factor of production?
   a) Land b) Human capital
   c) Management d) Labour

98. Which of the following cost is considered as returns from the missed alternative?
   a) MC b) Prime cost
   c) Opportunity cost d) Overhead cost

99. Crop farming and Milk production are examples of ________ enterprises
   a) Complementary b) Competitive
   c) Supplementary d) None

100. The law which is very useful in determining the best use of limited resources is
   a) Law of substitution b) Law of variable proportions
   c) Law of diminishing returns d) Law of opportunity cost

101. Which of the following is an economic factor affects both type and system of farming?
   a) Land b) Capital
   c) Labour d) Management

102. The costs of wages, salaries and payments for raw materials are the examples of
   a) Explicit costs b) Implicit costs
   c) Real costs d) All

103. The costs of self-owned and self-employed resources engaged in the farm are
   a) Explicit costs b) Real costs
   c) Implicit costs d) All

104. The total fixed cost, with increase in production

a) Decreases
b) Constant
c) Increases
d) None

105. The average fixed cost, with increase in production

a) Decreases
b) Increases
c) Constant
d) None

106. If the input is available free of cost to the farmer, he will operate

a) Anywhere in Stage I
b) Anywhere in Stage II
c) Anywhere in Stage III
d) At the end of Stage II

107. One can reduce the cost, by using more of the 'added resource', if

a) MRTS < Price ratio of factors
b) MRTS > Price ratio of factors
c) MRTS < Price ratio of products
d) MRTS> Price ratio of products

108. The problem of enterprises combination will arise only if the enterprises have relationship

a) Joint
b) Complementary
c) Competitive
d) Supplementary

109. The problem of enterprises combination will not arise if the enterprises have relationship

a) Joint
b) Complementary
c) Competitive
d) Supplementary

110. Decision making regarding the production level of products is insignificant, if the products have relationship

a) Complementary
b) Joint
c) Competitive
d) Supplementary

111. The profit maximizing criteria of a business firm

a) AC=AR
b) TC=TR
c) MC-MR
d) TC=AR

112. When TPP is maximum, MPP is

a) Equal to zero
b) Equal to APP
c) Maximum
d) Negative

113. Which of the following statements is correct?

a) When AC is increasing, it is more than MC
b) When AC is decreasing, it is more than MC
c) When AC is decreasing, it is less than MC
d) MC curve cuts AC curve at its top most point

114. If factor substitution ratio is equal their price ratio, then cost will be
   a) Increasing   b) Decreasing
   c) Maximum   d) Minimum

115. If factor substitution ratio is less than their price ratio, then cost will be
   a) Increasing   b) Decreasing
   c) Maximum   d) Minimum

116. If factor substitution ratio is more than their price ratio, then cost will be
   a) Increasing   b) Decreasing
   c) Maximum   d) Minimum

117. Which of the following is not a component of farm business?
   a) Land   b) Water
   c) Capital and Management   d) Labour

118. Average cost is
   a) Total cost/output   b) Fixed cost/output
   c) Total cost/ input   d) Fixed cost/input

119. APP is equal to MPP, when
   a) APP is minimum   b) MPF is maximum
   c) APP is maximum   d) MPF is increasing

120. Expenditure on fertilizers falls under
   a) Fixed cost   b) Marginal cost
   c) Variable cost   d) Both 'a' and 'b'

121. The costs which are not paid to others are called
   a) Fixed costs   b) Explicit costs
   c) Implicit costs   d) Overhead costs

122. Isoquant touches the axes in case of
   a) Power function   b) Exponential function
   c) Quadratic function   d) None

123. Farm management treats every farm unit unique in available resources, problems and potentialities because it has
   a) Broader field   b) Micro approach
   c) Practical field   d) None

124. Which of the following is a farming system on the basis of land use and culture practices classification
   a) Tenant farming   b) Specialized farming
   c) Mixed farming   d) Ranching

Answer Key

1	**D**	32	**A**	63	**A**	94	**A**
2	**A**	33	**A**	64	**B**	95	**C**
3		34	**D**	65	**B**	96	**C**
4	**B**	35	**D**	66	**C**	97	**D**
5	**C**	36	**D**	67	**B**	98	**B**
6	**A**	37	**C**	68	**B**	99	**A**
7	**B**	38	**A**	69	**A**	100	**D**
8	**D**	39	**C**	70	**B**	101	**C**
9	**A**	40	**B**	71	**B**	102	**A**
10	**C**	41	**B**	72	**D**	103	**B**
11	**B**	42	**A**	73	**B**	104	**C**
12	**A**	43	**A**	74	**C**	105	**A**
13	**C**	44	**B**	75	**A**	106	**D**
14	**A**	45	**A**	76	**B**	107	**B**
15	**B**	46	**A**	77	**A**	108	**C**
16	**C**	47	**A**	78	**B**	109	**D**
17	**D**	48	**A**	79	**A**	110	**A**
18	**B**	49	**B**	80	**B**	111	**C**
19	**A**	50	**C**	81	**C**	112	**A**
20	**C**	51	**A**	82	**A**	113	**B**
21	**B**	52	**B**	83	**A**	114	**D**
22	**C**	53	**A**	84	**A**	115	**B**
23	**B**	54	**A**	85	**B**	116	**A**
24	**B**	55	**C**	86	**C**	117	**B**
25	**A**	56	**C**	87	**B**	118	**A**
26	**C**	57	**C**	88	**C**	119	**B**
27	**B**	58	**D**	89	**C**	120	**B**
28	**C**	59	**A**	90	**A**	121	**B**
29	**C**	60	**C**	91	**A**	122	**D**
30	**C**	61	**D**	92	**B**	123	**B**
31	**A**	62	**C**	93	**C**	124	**A**

References

Adams, R.L., Farm Management, McGraw-Hill Book Company, Inc., New York, 1921.

Allan, N. Rae, Crop Management Economics, Granada Publishing Ltd, London, 1977.

Barnard, C.S. and J.S. Nix., Farm Planning and Control, Cambridge University Press, London, 1979.

Bishop, CE and W.D. Toussaint., Introduction to Agricultural Economic Analysis, John Wiley & Sons, Inc., New York, 1958.

Blagburn, C.H., Farm Planning and Management, Longman's Green & Co. Ltd., London, 1961.

Buckett, M., An Introduction to Farm Organization and Management, Pergamon Press New York, 1981.

Castle, EN. and M.H. Becker.. Farm Business Management: The Decision Making Process, The Macmillan Company, New York, 1965.

Christopher Ritson, Agricultural Economics: Principles and Policy, Granada Publishing Ltd, London, 1977.

Heady, Earl O, and J.L. Dillon, Agricultural Production Functions, Kalyani Publishers, New Delhi, 1960.

Debertin David, L. Agricultural Production Economics, Macmillan Publishing Company, New York, 1986.

Dhondyal, S.P., Farm Management (an economic analysis), Friends Publications.

Efferson, J. Norman, Principles of Farm Management, McGraw-Hill Book Company.

Forster, G.W., Farm Organization and Management, Prentice-Hall, Inc., New York, Inc., New York, 1953.

Handerson, James, M. and Richard E. Quandt, Microeconomic Theory: A Mathematical Approach, McGraw-Hill Book Company, New York, 1980.

Heady, Earl O., Economics of Agricultural Production and Resource Use, Prentice Hall, Inc., Englewood Cliffs NJ., 1952.

Heady, Earl O. and Jensen Harald, R., Farm Management Economics, Prentice-Hall of India (Pvt. Ltd.), New Delhi, 1964.

Hopkins, John. A. and Murray William, G., Elements of Farm Management, Prentice-Hall Inc., Englewood cliffs, N.J., 1955.

Johl, S.S. and T.R. Kapur, Fundamentals of Farm Business Management, Kalyani Publishers, Ludhiana, 1987.

Johnston, J., Econometric Methods, McGraw-Hill International Editions, 1984.

Julia H., Applications of Econometrics, Heritage Publishers, New Delhi, 1988.

Kahlon, A.S. and Karam Singh, Economics of Farm Management in India: Theory and Practice, Allied Publishers Private Ltd., New Delhi, 1980.

Klein, LR, An Introduction to Econometrics, Prentice-Hall, Englewood Cliffs, NJ.1962.

Maddala, G.S., Econometrics, McGraw-Hill Kogakusha Ltd., Tokyo, 1977.

Olson, K. and Westra, J. The economics of Farm Management : A global perspective. Taylor and francis, 2022.

Paul Loomba, N., Linear Programming: An Introductory Analysis, Tata McGraw-Hill Publishing Company Ltd., New Delhi, 1989.

Prem Kumar Gupta. and D.S. Hira., Operations Research, S. Chand & Company Ltd., New Delhi, 2000.

Robert O' Connor, Principles of Farm Business Analysis and Management, Irish University press, Shannon, Ireland, 1973.

Robertson, An Introduction to Agricultural Production Economics, Tata McGraw-Hill Publishing Company Ltd., New Delhi, 1971.

Rout, R.K. A text book of Farm Management, production and Resource Economics. Kalyani Publisher, 2021.

Sankhayan, P.L, Introduction to Farm Management, Tata McGraw-Hill Publishing Company Ltd., New Delhi, 1983.

Shankayan, P.L, Introduction to the economics of Agricultural Production, Prentice Hall of India Pvt. Ltd. 1988.

Sharma, A.N. and V.K. Sharma, Elements of Farm Management, Prentice-Hall of India Pvt. Ltd., New Delhi, 1981.

Singh, LJ., Elements of Farm Management Economics, Affiliated East-West Press Private Ltd., New Delhi, 1977 TahaHamdy, A., Operations Research: An Introduction, Macmillan Publishing Company, New York, 1989.

Subba Reddy S., P.S. Ram, T.V.N. Sastry, I.B. Devi, Agricultural Economics. Oxford & IBH Publishing Co.Pvt.Ltd. New Delhi, 2015.

Supriya, Mishra, P., Meena, N.R., Sharma, G., Choudhari, H.P.S and Meena, K.C. Agri-business: Management the fundamental of decision making, Jaya Publishing house, 2022.

Supriya, Mishra, P. Gangwar, R.R., Choudhary, S., Vani, G.K. Farm Management and its Practical utility. Jaya Publication House, 2018.

Supriya, Srivastava, A.B. and Mishra, H. Agricultural Finance and Cooperation Manual cum workbook. Narendra Publishing House, 2023.

William J. Baumol., Economic Theory and Operations Analysis, Prentice-Hall of India Pvt. Ltd., New Delhi, 1978. Yang, W.Y, Methods of Farm Management Investigations, Food and Agriculture Organisation of the United Nations, Rome, Italy, 1971.

Yang, W.Y., Methods of Farm management Investigation, Food and Agriculture Organisation of the United Nation, Rome.